East of
the Mountains
of the Moon

East of
the Mountains
of the Moon

Chimpanzee Society in the African Rain Forest

MICHAEL P. GHIGLIERI

THE FREE PRESS
A Division of Macmillan, Inc.
NEW YORK
Collier Macmillan Publishers
LONDON

The Free Press
A Division of Macmillan, Inc.
866 Third Avenue, New York, N.Y. 10022

Collier Macmillan Canada, Inc.

Printed in the United States of America

printing number
1 2 3 4 5 6 7 8 9 10

Library of Congress Cataloging-in-Publication Data

Ghiglieri, Michael Patrick
  East of the Mountains of the Moon.

  Bibliography: p.
  Includes index.
    1. Chimpanzees—Behavior.   2. Social behavior in
animals.   3. Animal societies.   4. Mammals—Behav-
or.   5. Mammals—Uganda—Kibale Forest Reserve.
I. Title.
QL737.P96G55   1988   599.88′44045′248   87-21203
ISBN 0-02-911580-9

To Conan Michael Ghiglieri
for his long wait

Contents

Contents

Illustrations

Preface

As a rite of passage into the ranks of field ecologists I spent two years in Kibale Forest, Uganda, studying wild chimpanzees in a project generally considered impossible. Though I went to Kibale with specific theoretical goals in mind, the apes and the forest soon forced me to broaden my narrow perspective. My experiences with the forest apes not only revealed a lot about wild chimpanzees, they expanded our knowledge of ecology and social structure among both chimps and man. *East of the Mountains of the Moon* describes my quest for the secrets the chimpanzees ultimately revealed: why chimpanzees live together as they do in the wild. It is a case study in sociobiology. But this book is also about an attempt to make contact with our closest evolutionary cousins, and it is about today's Africa, the tropical forest, and—because the tidal wave of human civilization is rushing ever faster over the vanishing world of chimpanzees—this is also a book about conservation . . . and about human values.

Some readers may be aware that this is the second book from my research in Uganda. My earlier book, *The Chimpanzees of Kibale Forest,* is a scientific monograph, loaded with terse statistical analysis of hypothetical models of evolutionary ecology intended for the professional and written with the barest minimum of accompanying detail. *East of the Mountains of the Moon* is not just a companion volume, it is the full account of the apes of Ngogo and Kanyawara as I have always hoped to tell it. It is not aimed at scientists, but it contains more science than most popular books on natural history because the adventure of the quest would be anticlimactic without the science of discovery at its end. *East of the Mountains of the Moon* is organized chronologically. But within this chronology several chapters also explore themes of their own, tell semi-independent stories, and, in places, break chronology and borrow from the future to fulfill their themes. I have tried to make such cases obvious and unconfusing. Several chapters also compare the chimpanzees of Kibale Forest with chimps observed in other areas to present a state-of-the-art look at our evolutionary cousins.

By the time I started fieldwork on the apes of Ngogo, well over one hundred person-years had been invested in studying free-living chimpanzees. These studies generally fit into one of two categories: long-term projects that provisioned the chimpanzees to allay their fear of humans, and much shorter projects that attempted to study the apes with an absolute minimum of disturbance. The potential I saw in the latter studies convinced me to attempt my project the way I did. But the theoretical rationale for my work was drawn from the two long-term projects in Tanzania at Gombe Stream and Mahale Mountains National Parks. I owe a tremendous debt to *all* those pioneers of chimpanzee behavior and ecology, especially to the work of Jane Goodall and Toshisada Nishida. And I tip my hat to the intrepid Japanese scientists who routinely trekked many miles for small returns in data from skittish apes. As mentioned above, I compare my observations with those of this diverse fraternity of scientists intrigued by chimpanzees, and, for the convenience of the curious reader, I have identified the sources of their more pivotal reports in the text and also listed them in the reference section at the end of this book.

Acknowledgments

Many people made this work possible or assisted in its completion, and I owe thanks to all of them.

I could not have conducted research in Kibale Forest Reserve without the permission and helpful assistance of Mr. Frederick B. N. Mukiibi and Mr. X. K. Ovon of the National Research Council of the President's Office, of Mr. Peter Karani of the forestry department of Uganda, and of Professor Ramsis Lutfy and Mr. Simeon Semakula of the department of zoology, and Mr. Anthony Katende of the department of botany of Makerere University. Thomas T. Struhsaker and Peter S. Rodman were instrumental in making me aware of the potential for research on chimpanzees in Kibale. The National Institutes of Health (through grant number MH 23008-04 to T. T. Struhsaker), the Boise Fund and Professor G. Ainsworth Harrison of Oxford University, the U.S. National Academy of Sciences, and the New York Zoological Society supported portions of my research expenses and data analysis. I am extremely grateful to all of them.

Davis Semikole eased my entry into Uganda and provided orientation that probably saved me from several snags. Simon Wallis introduced me to forest botany invaluable to my research.

Acknowledgments

Steven Mugisa Yongili, in addition to maintaining camp and supervising the thankless and unending task of keeping nearly sixty miles of trails and track cut, at my request provided me with information about local politics and illegal incursions into the forest, occasionally at some risk to himself. Ben Alfred Otim deserves the highest commendation for his allegiance and dedication to the position of game guard—a job paying as poorly as anyone can imagine, yet demanding integrity that, if possessed by our politicians, would transform them into effective leaders. Otim consistently performed his duty in the face of adversity.

Tom Struhsaker was not only instrumental in connecting me with the apes of Ngogo, he was consistently interested in my research and offered many useful suggestions to improve my collection of data and their analysis. Lysa Leland's friendship, hospitality, cheerfulness, and good cooking during my stays with them in Kanyawara meant more to me than she will ever know. Both Tom and Lysa went out of their way to enrich life for me. I thank them both.

Joe Skorupa and Lynn Isbell were unfailingly cheerful in providing me with meals and space to sleep when I found myself stranded in Kanyawara for lack of petrol. Despite their difficult research schedules, they always made time to help me out, and consistently burned plenty of rice for dinner. I have never felt more welcome anywhere.

Visits to Fort Portal generally were an unpleasant routine—it is the only place I have been seriously threatened with decapitation by a high government official. But the Missionary Sisters of the Holy Cross—Sisters Patricia, Gabe, Edward Ann, Jeanne, and Catherine Ann—offered and delivered boundless hospitality that often transformed that routine into something special. Brother Tom and Father George also shared their friendship, home, and board with me and enriched my stay in western Uganda.

During Amin's regime, Kampala was probably the least agreeable place on the continent. But logistics and official duties sometimes dictated that I be there. I can never thank Oskar and Linda Rothen of the Swiss Consulate enough for their limitless hospitality given with never a question asked. Beyond their much appreciated hospitality, they offered sanctuary.

Most supplies of the sort that do not grow in a garden or cannot be carved from a tree were not available in western

Acknowledgments

Uganda during my research. My parents, Lloyd and Fern Ghiglieri, not only watched over my affairs in the U.S.A., they also promptly procured and mailed supplies I requested and luxuries that I did not. I wish to thank them, a thing I do too infrequently, and to thank my friends Debra Suzan Judge, who took time from her own research in Baja California, and Linda Scott, who did the same in Gilgil, Kenya, to send me logistical supplies and reminders of our common culture.

Again, I never could have conducted this research with the scientific goals I had in mind were it not for the many scientists who conducted research on chimpanzees prior to (and during) my own. My debt to them is incalculable.

I wish also to express my gratitude to the Ugandan people for their friendship and for assistance and hospitality whose equal I have yet to find in any other country.

Last, but foremost, I wish to thank Conan Michael Ghiglieri for understanding my two difficult years of absence.

Various drafts of the text were proofread and criticized helpfully by Laurie Benz, Roger Billica, Constance Ghiglieri, Lynn Isbell, Debra S. Judge, Christian Kallen, Lysa Leland, Richard Shapiro, Joe Skorupa, and Karen Schmidt, who went beyond the call of friendship with some very valuable critical analysis. Without a doubt their inputs improved the final draft, and I am indebted to all of them, though I claim full responsibility for any shortcomings which may remain. Constance Ghiglieri not only helped with most of the drafts of this book and created good working environments in which I wrote, she also worked as my field assistant during my final two months at Ngogo. I owe her thanks many times over. I also would like to thank John Brockman, Katinka Matson, and Edward Rothstein for having faith in this book and Ed Rothstein again for his valuable insights in directing it into a coherent story.

East of
the Mountains
of the Moon

— 1 —

Killer Apes

I lifted my field glasses again to ascertain exactly which part of Gray's body Zira was grooming. These days Gray, an ancient and wrinkled female scarred by battles even she had probably forgotten, spent about half her time with Zira, a sleek young adult female who presented a total contrast with Gray's been-there-and-done-that persona. The two were close. Closer than any other two adults I observed in Kibale Forest. And even though I had known them for only six months, every time I found them visiting some giant fruit tree in the forest at Ngogo I found myself wondering again if Gray was Zira's mother. If she was not, the only other plausible basis for their mutual attraction was friendship.

In chimpanzee society friendships are usually strongest between relatives, but familial solidarity—not "friendship"—really is the more accurate term to describe such positive affiliations. "Friendship" is meant more to refer to relationships between unrelated individuals. But because female chimpanzees normally emigrate from their mother's social group at maturity and may never see their mothers again, solidarity between adult females has been observed extremely rarely—so rarely as to be considered

1

negligible. If Gray and Zira were mother and daughter, they were an unusual adult pair—and if they were unrelated friends, they were an even greater anomaly. Gray and Zira were a mystery, but to me not a trivial one, and I never stopped watching for clues to solve it.

Zira was grooming Gray's back. Both apes sat on a huge horizontal limb of a giant tree emerging from the canopy of the forest and leaning over a steep ravine cloaked with a jungle of vegetation. A fall would have killed either of them, but they appeared as at ease as if they were sitting on the ground. Gray sat resting her arms on her knees, and Zira sat immediately behind her as if riding second in a crowded dugout canoe. But instead of gazing at the scenery, Zira's nose was only a finger length away from Gray's back as she used both hands to part the older female's light colored hairs. Zira scrutinzed Gray's skin as if it were vital that no parasite or particle of foreign material escape detection and removal. As were most chimpanzees, Zira was serious about grooming.

Zira stopped grooming abruptly and turned around on the limb to now sit back to back with Gray. As plainly as if Zira had announced verbally that it was time to switch roles, Gray slowly sat up, blinked as if snapping out of a dream, then turned to sit facing Zira's back. Gray smacked her lips with audible pops as she leaned toward Zira. With exaggerated care she parted Zira's glossy hairs with her left hand (Gray had lost a finger from her right hand and rarely used it for manipulating things) and searched Zira's skin for foreign matter.

A great blue turaco flashed across the sky between the apes and me. I stood in a patch of low growth thick with new saplings struggling against one another to shoot up first and monopolize the sunlight. This was the best spot for viewing Gray and Zira. I was about forty yards from them now but I had been a lot closer during most of the past three hours of their gorging on wild figs and lolling about waiting for enough appetite to eat again. Inexplicably Gray had accepted me very quickly after we had first met. By contrast, Zira had fled from me in near panic. But Gray's insouciance had eventually converted Zira so that now she too virtually ignored me.

I glanced again through my field glasses at Gray's utter absorption with grooming, then I jotted notes on what part of Zira she was grooming and for how long. My notes were beginning to accumulate, and the morning had been idyllic—almost too good to be true. Then, suddenly and sharply, maybe fifty yards away, a limb cracked. Though it was still hidden, I knew what animal had snapped that limb. I immediately looked for a tree to climb, but nothing other than saplings grew near my vantage spot. Upslope the tangle of vegetation vibrated, jerked, and snapped in two distinct places as the invisible creatures advanced rapidly toward me. Through the obscuring foliage I caught glimpses of huge beasts coated with dried red mud. Elephants. Only seconds away. And stampeding directly toward me.

The early 1960s saw a new wave of long-term field studies of nonhuman primates in the tropical regions of the earth. Funding agencies were assured that anything we learned about the behavior and ecology of our relatives, the primates, would better our understanding of ourselves and our antecedents. In general, each field study was more refined and sophisticated than its predecessor. Researchers began asking specific theoretical questions about wild primates, and some were receiving clear answers.

Early in this phase of primatology, though, it became clear that our savanna-dwelling relatives—the baboons, vervets, and patas monkeys—had little in common ecologically with early man. None of them subsisted on a big game economy, as was postulated for our antecedents, the early australopithecines and *Homo habilis*. Unfortunately for science, we had no evolutionary cousins on the postulated savanna road to humanity to show us how it had been for our own ancestors. Some scientists concluded that we would learn more by studying closely related species living in forested habitats, even if only to show us the evolutionary results of having adapted to alternate ecological situations. And many of us turned away from savanna primates to focus on other species related to us more closely, specifically the great apes.

Genetically chimpanzees are our closest living relatives. Numerous biochemical comparisons have confirmed this, culminating in 1984 in actual genetic matches of DNA's by Charles Sibley and Jon Ahlquist revealing that we differ by a mere 1.5 to 2 percent. In the parlance of taxonomy, we and chimpanzees are sibling species. And as observations of chimpanzees by Jane

3

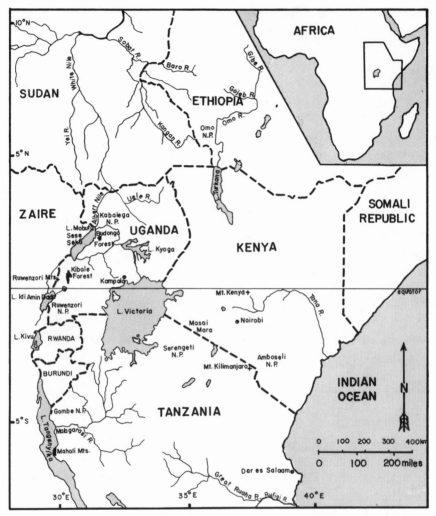

EAST AFRICA

Goodall and her colleagues at Gombe Stream National Park and by Toshisada Nishida and his colleagues at Mahale Mountains National Park in Tanzania accumulated, the apes unanticipated similarities to ourselves repeatedly demonstrated that our definition of man was inadequate. Chimpanzees not only used tools, they constructed them in a culturally transmitted way, often in advance of immediate need. They not only hunted large mam-

4

mals, they often shared the prize as well. Chimps in laboratories not only learned hundreds of human words in American Sign Language (ASL) and other artifical languages, they sometimes invented words they needed but had not been taught. Some chimpanzees, notably those cared for by Roger and Deborah Fouts, are themselves teaching American Sign Language to untutored chimps. Last, and most disturbing to people believing that *Homo sapiens* had a monopoly on cold-blooded murder, in the 1970s Goodall and several of her colleagues observed bands of male chimps at Gombe apparently systematically patrolling the fringes of their territory and attacking and killing males from an adjacent territory.

Why were chimpanzees killers of their own kind?

This question, and the broader questions of what determines social systems, drove me to Africa.

Originally my plan was to start yet another study of chimpanzees at Gombe, where killings were known to be taking place. But in May 1975, when I was still trying to line up grant support for my work, my project became impossible. In a daring midnight raid, a platoon of about forty guerillas of the Marxist Popular Revolutionary party motored across Lake Tanganyika from Zaire and infiltrated the main study camp at Gombe. After searching in vain for Jane Goodall, they grabbed a Dutch research assistant and three students from Stanford University before returning to Zaire. The release of one student, Barbara Smuts, with ransom demands prompted negotiations between the rebels and the hostages operatives: Stanford, the U.S.A., and Tanzania. Although the hostages were released unharmed after the ransom was paid, worries over additional scandals of international extortion prompted Tanzania to close Gombe to long-term visits by non-Tanzanians.

My plans became just one of several scientific casualties of this midnight raid. But I was still determined to investigate the ecological aspects of chimpanzee life in order to explore whether limited resources affected social life and, as much as possible, to investigate whether lethal warfare between groups was normal or an anomaly. This now was literally the hottest topic in the science of animal behavior, one rattling with contention over the merits of controversial sociobiological explanations versus more traditional arguments based on the premise that individuals are programed by nature not to act against the good of their species. But, hot or not, I was in a quandary as to how and where I could conduct an investigation that would test these arguments. Peter S. Rodman, my major professor at the University of California, Davis, came to my rescue by introducing me to Thomas T. Struhsaker, a primatologist of high repute working in Uganda. In response to my letter Struhsaker wrote, "The chimps are not so tame nor hominized as the Gombe chimps seem to be, but I believe they could be studied."

Struhsaker's letter convinced me that Uganda's Kibale Forest offered the solution to my problem. But during the whole of 1976 all the funding agencies I contacted to support my work agreed with one another that the project was impossible, or close to it, and they added that even if it were possible, Uganda was too dangerous a place in which to attempt it. The unpredictable and murderous reign of Idi Amin Dada was routinely grabbing spots for Uganda in the international news. Even so, Struhsaker's six years of uninterrupted research seemed to show that African politics had little bearing on science. Finally, despite having gained no financial support, in December of 1976 I took Tom up on his offer of an airfare pirated from his own grant plus shared use of an old Land Rover. I sold my car and liquidated what little else I could, adding the proceeds to my savings and my educational benefits for Vietnam era veterans. While not posh, my prospects looked reasonably good.

The landing gears thudded against the tarmac. I joined the group of Africans filing out of the jet at Entebbe International Airport. Tom Struhsaker, stationed on the opposite side of the country in Kibale Forest, had asked one of his students at Makerere University in Kampala, Davis Semikole, to meet me. Semikole not only guided me for three days through an unfamiliar bureaucratic jungle which would have been infinitely more difficult without his assistance, he was even en route to Kibale Forest for a field class Tom was teaching. After I was equipped with a pupil's pass, a student visa, a Forest Department research permit, a National Research Council identity card, and a fistful of letters of introduction to the police commissioner and district commissioner of Toro District (in which lay Kibale For-

6

est), to Makerere University, to whom it may concern, et cetera, Semikole and I were free to find our way to Kibale Forest.

We trekked for nearly an hour from the university, me humping a seventy-five pound pack, to reach the buspark downtown. Kibale Forest was 200 miles west of Kampala, and the only transportation available consisted of unreliable buses. Because in Uganda spare parts had nearly dried up due to the lack of hard currency for their importation, buses were, in fact, dwindling in number as well as reliability. We sat all afternoon and evening in the government buspark, in hopes that the promises were true that the bus to Toro District would be leaving that day. Around midnight the officials behind the window admitted the bus would not show up at all, so we spent the night in a cheap hotel, but arose before dawn to hurry to a privately run buspark on the chance that they would have a bus going.

Not long after we arrived a man shouted for everyone interested in going to Toro District to line up to buy tickets. Competition for seats was stiff; everyone stampeded for a place at the head of the line. While I marveled that no one had been seriously injured, the manager called me to the head of the line.

"You are going to Fort Portal?" asked the driver's assistant.

"Yes. Is this the right bus?"

"Of course. But we are not having room for so many people. So you can buy your ticket first."

"Why should I be ahead of these others?"

"We can see that you are not from Kampala. We don't want you to be stranded here. It can be a bad place. How many tickets do you need?"

"Two. . . . Thanks very much."

"Not so much."

I was impressed by this favor, but soon learned that this solicitude for a visitor was a quality of many Ugandans.

The bus was crammed with people, bananas, chickens, and chatter. Although progress along the muddy road was slow and halts frequent, we rolled past undulating, verdant hills that seemed to extend indefinitely. We also rolled past the government bus that had not arrived from Toro, broken down and leaning precariously in a muddy ditch. Grass-roofed, mud-walled huts and groves of bananas dotted the landscape. Uganda nestles on the fringe of the zoological paradise of East Africa and joins the savanna and its incomparable big game with the mysterious jungles of Zaire in the region of the fabled Mountains of the Moon. I watched for some of this big game, but goats and a few zebu cattle were all I saw. A young herdsman with a spear was the main reassurance that I was on the right continent.

We finished a large hand of bananas and then gazed again out the window. The seats were too close together and the jolting of the overloaded bus was rearranging my kneecaps. Kibale Forest seemed close now. Doubts about my being able to make the breakthroughs necessary with the chimpanzees to test the hypotheses about what determines social structure crept into my head again and demanded attention. This issue, especially regarding systematic killings by the apes, could not be resolved without more data; as things stood now no consensus existed as an explanation. But I thought I had it.

After Jane Goodall's research team cut down on feeding bananas to the Gombe chimps, whom she had decided could not be studied initially without such provisioning, about a dozen of them drifted south to the Kahama Valley and were rarely seen again near camp. Thirty or so members of the larger community remained near the Kasakela Valley and continued to visit camp for occasional bananas. By weaning chimpanzees who had been living a semiwelfare life-style of easy food, the researchers expected to reinforce the apes old habits of searching the woods for food. In the next few years Goodall's team discovered bands of males traveling together "scouting" on what were termed "patrols." Then, one after another, these patrols of Kasakela males were observed to systematically catch nearly every Kahama male when he was with only few companions and kill him. Over the span of a few years, all seven males vanished, most of them as observed victims of fatal gang attacks. The murderers and victims who once had eaten bananas together at the provisioning station became combatants in a war that annihilated the small Kahama Community.

Why? The answer to this important question could not be attained simply by reconstructing the picture at Gombe; too many other questions remained unanswered. For instance, I suspected an ecological pinch, namely the loss of massive quantities of bananas that had been reliably available for many years had created a competitive atmosphere, triggering a competitive behavior of which the apes already were capable, specifically killing. But because the killings occurred during an extended period when few estrous females (adult females in heat) were available, however, Goodall suspected the Kasakela males were seeking mates rather than food. Both of us were guessing. And to complicate matters, researchers twice witnessed Kasakela males attacking strange females in the north and brutally murdering the infant of each. The bottom line is this: the Kasakela males who murdered at least seven of their southern neighbors gained more of the two main resources which limit their reproductive success: adult females and an expanded territory containing additional food for themselves and their offspring. How can one know whether either or both of these explanations hold water?

My project in Kibale relied on the following logic: If male chimpanzees kill alien male competitors as a natural part of their social system (rather than as a result of an artificially created situation) to expand their territories for additional mates and food, then evidence of it should be apparent among other communities of wild chimpanzees. And, if killing is a natural though not necessarily frequent part of their repertoire, males should behave in other ways as if this were the case. I was hoping that the chimpanzees in Kibale would reveal enough of their lives to me to make this judgment statistically. Also by working in Kibale, a habitat more similar ecologically to the core range of chimps as a species, I would actually improve my original plan of trying a nonprovisioning study of a new community of Gombe chimps, because Kibale should be more representative of the "average" chimpanzee. The main difficulty I anticipated would come from not being able to feed the chimps to get them to accept me. Previous studies suggested this would be an almost insurmountable barrier.

Both earlier long-term projects in Tanzania relied on feeding the chimpanzees—up to 600 bananas per day at Gombe, enough to supply 40 percent of the caloric demands of the entire community—to allay their fear of humans. Researchers in both places assumed, perhaps rightly, that the apes would otherwise never allow close observation. But provisioning was not the only route that had been tried. Even before Goodall arrived at Gombe, Adriaan Kortlandt (1962), a Dutch pioneer in protohominid research, was observing chimpanzees in Guinea from behind blinds, which, while a successful method, was one inherently too self-limiting for an ecological study conducted at large in a forest. A clue that both of these methods might be circumvented was provided by a pilot study conducted productively for half a year by Vernon and Francis Reynolds in Uganda's Budongo Forest. They had loosely adapted the technique pioneered in 1960 by George Schaller on mountain gorillas: carefully edging closer to the apes each time he found them while copying their own inoffensive communications and behaviors. But despite the promising start by the Reynolds, most experts still agreed that this was impossible with chimpanzees.

Why was it out of the question for me to provision chimpanzees also, just to get my project rolling? Evidence from reports in the 1970s of the Tanzanian projects by several researchers, Toshisada Nishida, Curt Busse, Anne Pusey, David Riss, Richard Wrangham, and Vernon Reynolds, indicated that feeding the apes was changing their ecology, altering their patterns of travel, and even distorting their basic social behavior. Providing food not only complicated the interpretation of the naturalness of apes killing one another, it left me with no option but to make George Schaller's technique work.

Strangely, it is not immediately apparent how merely feeding a group of animals can have so many side effects. The dogmatic explanation is basically Darwinian. It maintains that ecology, or the abundance and dispersion of environmental resources, constitute one of the primary dimensions of natural selection. Because over the lifespan of every animal resources critical for its survival are limited, competition exists. And because this competition is most intense between members of the same species, ecology has an inordinate influence on social behavior. Consequently, a prolonged distortion of the natural ecological relationship between animals and their environment can readily distort their social relationships with one another—in the case of providing food, by reducing overall competition. And conversely, in the case of removing food, by increasing competition. Such is the scientific logic. But so many people found this logic too theoretical that I eventually found I had better success in explaining it by using the following analogy with humans in place of chimpanzees.

Imagine a starship winking out of hyperspace near the third planet from the sun. In it are ourselves, a team of extraterrestrial scientists who have traversed hundreds of light years through the depths of space. Through the viewport we gaze upon the earth, a blue-green jewel suspended in emptiness. Our assignment is to learn everything we can about its intelligent life.

At random we choose a dry region about a third of the way from the southern tip of a continent straddling the equator. As our ship hums into the atmosphere, we wonder what we will encounter below. We land on the edge of a temporary lake in the Okavango Delta on the northern edge of the Kalahari Desert.

We watch some of the intelligent inhabitants of the planet gathering outside. They seem curious. They are odd-looking creatures with black growths on their heads. They are not frightening; they are only half our size and look like primitive versions of ourselves. We step out through the portal. The earthlings take one look at us and flee.

Months pass and they still do not trust us. Whenever they see us they flee as if we were monsters. So we try hiding and are able to make limited observations of their habits. But we need more. We do find that they eat many different foods but depend on two important ones: the large nut of a common tree and the flesh of four-footed mammals. We decide to offer some of these foods as gifts near their camp: perhaps we can attract them to us. With the synthesizer aboard ship we can produce any amount of the stuff.

It worked. One of the males now visits our free food area regularly. And occasionally one or another has come with him. By sitting nearby we manage to communicate our friendliness to this male. We hope that soon he will become habituated and act as if we were not present.

Six months pass and the majority of those living in this camp now regularly visit our provisioning area. We have observed that the males have the prerogative of eating first, but even among them there exists a dominance hierarchy. Once, when the synthesizer broke down, we were unable to provide the daily feast of what these natives call mongongo nuts and meat. The females squabbled with one another and the males became very rough. At last one of our technicians repaired the synthesizer and we hurried out with food. They fell on it like a pack of beasts; the males pushed and shoved one another, and the females begged pitiably while their children cried and begged also.

A year has passed. Rarely do males or females leave camp to find food. Why should they search the scorching desert when they can get food from us? They still rob an occasional nest for eggs and gather other rare foods, but we provide them with nearly half of what they eat. Our excellent observations have revealed how quarrelsome they are—lazy too. But their children are so cute that we laugh despite ourselves; it is all we can do to keep ourselves from playing with them.

After seven years we have learned a great deal about how these earthlings behave and how they interact with one another. But our picture of how they would make their living naturally is not clear. Because of all this free food, they live a life of ease. We decide to cut down on the feeding in the hopes that they will revert to their natural habits of hunting and gathering so we can determine how their social struture, ecology, and technology are related. The years are nothing to us but they might find readjustment to their old ways difficult after so long. Many of the children have no memory of their parents seriously foraging for food. You say we should have cut back a long time ago, but hindsight is easy. We will still feed them every ten days to keep them visiting so we will not have to search for them.

In the absence of free food they have returned to the desert. They look thinner, but life must be much as it was before we came. A few have moved to a small camp not far from here. Recently some of the males raided this camp: already they have stolen from and killed males who once feasted on our synthetic mongongo nuts and meat here in the provisioning area. We wonder now what is normal for these earthlings. Did they routinely war on one another before we arrived, when food was harder to get—or do they kill now for the first time only because times suddenly have become difficult?

Lest my analogy be taken as derogatory with regard to the earlier work on chimps in Tanzania by Goodall, Nishida, and their colleagues, I must emphatically point out that it is meant *only* to provide perspective. These were, and still are, amazing and valuable projects never to be repeated or matched. Projects worthy of great respect. By now I had read Goodall's *In the Shadow of Man* four times; that, combined with all her other reports and those of other chimp workers in Tanzania, were the inspiration for my project. They were responsible both for raising the questions concerning chimpanzees killing other chimpanzees and for providing a wealth of data about wild chimpanzees to help me with the design of my own research. The issue of how provisioning the Gombe chimps may have influenced their overall behavior is a question that has been asked of Jane Goodall innumerable times, and, without doubt, Goodall herself has provided the best answer, "Yes, feeding bananas probably has changed some of the chimpanzees' behaviours, but how much, and in what way, we still don't know."

The reason that question is unanswerable is due to the lack of extensive baseline data on chimpanzee behavior and ecology *prior* to banana provisioning. In other words, no data from a control group exist for comparison. One of the justifications for my research was to provide a partial remedy for this lack. At this time no report existed interrelating the ecology of unprovisioned wild chimpanzees with their social structure and analyzing the relationships between the two in sociobiological terms. Obviously I was convinced my project was a necessary next step. The main rationale for my being here was to investigate the factors causing the unique social system of our nearest living relatives. Among other things my project would be an exploration of the natural history of competition and selfishness and, by extension, survival.

I did have some preconceived notions to guide me in my quest. The first one explained killer apes as a natural and adaptive phenomenon. The first hurdle to understanding this is abandoning the old notion that each individual of a species is somehow programmed to act in the higher interest of perpetuating that species *even when it goes against his own reproductive interests.* By contrast, neo-Darwinism, based on genetics and observation, has revealed that in general individuals act to maximize their own reproductive success even at the expense of that of another individual. In fact, it is as if each individual is merely a machine programmed by its genes to reproduce as many more of the *same* genes as possible (in the form of offspring). British biologist W. D. Hamilton's insightful work on social insects reveals that each individual acts not only to reproduce its own genes, but also those being reproduced by close relatives. Hamilton coined the phrase "inclusive fitness" to describe the phenomenon in which individuals act in ways which assist near relatives to reproduce more successfully and thereby increase the reproduction of genes they share. Depending on circumstances this assistance could take on many forms of apparent altruism, including dying during defense. Inclusive fitness, I thought, was the key to understanding chimpanzees.

Researchers at both Mahale and Gombe found that, while it was very common for females to emigrate from the community in which they were born, males always remained in their natal territory. Not only is this nearly unique among nonhuman primates, it creates very unusual genetic consequences. Male chimps beget sons who beget inbred grandsons and so on for several generations with no new genes from males. Ultimately each chimpanzee community must contain males who share many more genes in common with one another than with males from other communities—or even with their own mates. Sharing more genes in common opens the evolutionary door for males to evolve cooperative behaviors which increase their *overall* reproductive success through inclusive fitness.

Although still theoretical, the logic of inclusive fitness makes it easy to see why males might cooperate in defending their communal territory: the territory holds everything vital for reproductive success. The Big Question is whether these reproductive advantages are great enough to make attempting to kill rival males worth the risk. The hint that they may be worth it is this: the additional territory gained by successful warfare holds more of the two most important resources vital for increased reproductive success, additional females and more food for additional offspring. And, because offspring are the currency of the bank of evolution, any behavior that increases offspring at the expense of competitors is going to be reinforced by natural selection.

Again, this was prevailing sociobiological theory, and, while heartless, it provided the most compelling logical framework for understanding how killing was natural. The rules it provided for competition between chimpanzee communities were anything but peaceful; they seemed more like Hitler's Nazis trying to conquer all of Europe for Lebensraum. But violent events, though significant, obviously must be rare. I knew it would be a fluke of probability for me to witness lethal warfare in Kibale Forest. And, because I knew I could not count on being able to observe lethal clashes, I planned on searching for answers indirectly. For instance, if this inclusive fitness hypothesis was correct, male chimpanzees in Kibale Forest would reveal a sharp preference for one another's company, and their general behavior toward one another would tend to cement solidarity between themselves at the expense of their relationships with females. There should be many ways to document these trends by observing and recording day-to-day behavior. To accomplish even this was a serious challenge. The experts were still betting that it was impossible.

Dusk bathed our rattling bus crimson as we rolled into the old colonial enclave of Fort Portal. Semikole began immediate negotiations with taxi drivers to hire a ride to the Kanyawara Forestry Station where Struhsaker lived. None of the drivers wanted to take the risk. But Semikole waxed eloquent and produced a fat wad of Ugandan shillings. The legal equivalent of fifty U.S. dollars convinced one to try.

More grass-roofed houses blurred past as the front wheels shimmied over potholes. Meanwhile, the driver and his two cronies in the front seat carried on a heated discussion in Ratoro. As if to perform a disagreeable duty, one of the three occasionally glanced through the windshield to ascertain that we were still on the dirt road. The driver suddenly stomped on the brake pedal, locked all four wheels, and we skidded to a halt only a heartbeat short of a young girl who stared at us with eyes and mouth in circles of astonishment. Statistically one is eleven times more likely to die per kilometer traveled in SubSaharan Africa than in North America. I was beginning to see why.

A half hour from Fort Portal we arrived at Kanyawara, in total darkness. The driver stopped as the trees thickened and refused to go on, saying, "This is the jungle. People *die* out here!"

Semikole whispered to me that the road ahead probably was impassable to the taxi anyway, so we paid off the driver and I curbed my disgust at his skills. The cloud-darkened forest was jet black. By feel we trekked up a muddy track for about ten minutes. I wondered what lay ahead, but I felt too tired to concentrate on imagining.

On the crest of the hill a kerosene lantern glowed dimly through the doorway of a small, isolated, corrugated iron house. Semikole shouted hello, and a tall, ascetic-looking white man with a flowing gray beard eclipsed the light. He was wearing a T-shirt and a native *kanga*, a length of colorful native cloth, wrapped around his waist. He looked like a guru for some LSD cult, but then, I probably looked like his lieutenant. I considered saying, "Dr. Struhsaker, I presume," but something held that cliché back.

"Ah, Semikole, *Habari gani?* (what news?) Tom greeted. "And Michael?"

"Yes. How are you? I'm glad to have finally arrived," I managed. It had taken me a full week to reach this isolated hilltop.

Lysa Leland, Tom's wife, appeared smiling from the mud-walled annex to the single-roomed corrugated iron shack. She greeted us enthusiastically and relieved me of my heavy pack. Then she showed me around, and made us feel at home with bowls of steaming curry.

Abruptly Lysa asked, "Do you hear them?"

I listened. From out of the inky blackness to the south came distant wails and eerie screams, a piercing rising cacophony of unearthly shrieks that made my hackles rise. It was the pant-hooting of chimpanzees, their long-range communication.

"They're welcoming you," she explained.

2

In the Forest

"Welcome to Uganda!" Simon Wallis pumped my hand. Tom had just dropped me off at the Ministry of Works (the government repair shop) in Fort Portal to rendezvous with Simon, who was having Tom's old Land Rover repaired. "It's a superb country," Simon continued. "Actually I've been back here only a week myself."

"Thanks. Where *have* you been?" I had known vaguely about Simon, a twenty-four-year-old graduate student at the University of London. I would be sharing both the Ngogo (pronounced en-go′-go) study area and Tom's old Land Rover with him during the last few months of his research on grey-cheeked mangabeys, an omnivorous social monkey.

"Home in the U.K. for the past two months. At the University." Simon looked thoughtful for a moment. "Tom must have told you."

"Told me what?" I was having one of those there's-something-I-don't-know feelings.

"About poachers burning me out at Ngogo two months ago. My hut, my data on the mangabeys, my cameras . . . *everything*. Anyway, I had sent carbon copies of most of my data home by post. But I had no copies of the maps so I lost all the ranging data. I figure in five more months I'll be able to replace most of them."

"No. He hadn't mentioned that. He did tell me I would love it there. He said he and Lysa considered their visits there almost vacations."

"That's Tom alright. But he's right: the forest is much better than at Kanyawara. Undisturbed, and not so many steep hills. I guarantee you'll be glad you're there instead. Also the trail system covers a much larger area—and there are more chimps."

"I hope so. I just spent two days in Tom's trail-grid system at Kanyawara and did not even find a chimp footprint. But I know they were around; I heard pant-hoots during most of my first night there." This had been a disappointment to me, as well as a suspected glimpse of the future, but I was still hoping Ngogo would be easier.

"I'm not surprised," Simon admitted with a distinct air of knowing he knew something I did not. "Chimps are not easy. Looks like the boys are done here." (He said this regarding two mechanics, both of whom were older than he was.) "How about tea at the Mountains of the Moon [Hotel] before we try the road?"

"Sure." The road, not tea, was what I wanted, but I figured I should begin this relationship agreeably.

We finally headed south along a dirt road kept in repair by Ministry of Works gangs armed with shovels. Ngogo, which would be my study area for two years, was more than thirty miles south of Fort Portal, and twenty-four miles by the roundabout road from Kanyawara. The drive from either place required about two hours. A foot route connected Ngogo with Kanyawara along the trail-grid system and elephant paths winding nine miles through the forest, but because of mud and swamps it was logistically inferior to the road.

"If Ngogo is so much better than Kanyawara, why isn't Tom based there?"

"I don't think Tom even knew about Ngogo during his first few years after he settled into Kanyawara." Simon answered, rubbing his scraggly beard with one hand. "There was no track to it then—it was only a temporary camp for poachers. Old Bill Freeland opened Ngogo as a study area in 1974 so he could observe mangabeys in a habitat different from the one at Kanyawara. By that time Tom had been there for four years. Tom did catch onto it quickly, though, and started doing samples on redtails and red colobus there for a week at a time. But Tom wouldn't fancy living at Ngogo: too far from town—uses too much petrol—and it's a bloody awful track." Simon braked abruptly for a herd of about twenty scrawny zebu cattle plodding along the road between the banks, which were walled by stands of bamboolike elephant grass at least ten feet tall.

"They look pretty thin," I commented.

"Yeah, but they never look better. European breeds don't survive here. This is not really cattle country; that's further south, Ankole District."

As we traveled further south fewer *shambas* (small family farms) and huts dotted the hills. The road snaked through ubiquitous stands of elephant grass punctuated by agricultural plots, including one giant tea plantation, then entered ever more frequent belts of forest.

Simon braked again, this time for a troop of olive baboons galloping across the road. The adult males looked as if they weighed at least seventy-five pounds.

"Look at those babs! Superb! That's what I came here to study sixteen months ago." He looked wistful.

"What went wrong?"

"Habituation. The same problem you will be facing with the chimps. Studying arboreal monkeys is not much trouble because they *are* so arboreal. To habituate a group you just follow it. They can't escape because they are slower in the trees than we are on the ground, so you just stay with them. When I habituated S.B. Group [of mangabeys] we followed them for ten days all day long from one sleeping tree to the next night's sleeping tree. Even before ten days they were habituated. But I spent my first two months trying to habituate the baboons. And every bloody time I got close they ran away. There was no way to keep up with them; they are too fast on the ground. I had to give it up. It was depressing. So I switched to the social behavior of F.F. Group [Freeland's mangabeys]."

An hour south of Fort Portal Simon steered us onto a side track just wide enough for the vehicle. Upon reaching the first bend he abruptly steered straight through a wall of elephant grass. Visibility was no greater than ten feet.

Elephant grass snapped and parted in two waves as if before the onslaught of some hell-bent mechanical harvester. Crashing blindly and uncontrollably through that cracking vegetation was not pleasant. We had to close the windows to avoid being slapped by the stalks. The cab heated up like a sauna. How long would this ride last? I studied Simon's profile. His long ponytail was littered with leaf fragments, but his poker face revealed no clue. I concluded, though, that his nonchalance was macho for my benefit.

A few hundred yards beyond we swished into another world and rumbled through its lush understory, the lowest thirty feet of herbs and small trees in the forest. Leaf-shaped, tan grasshoppers and a variety of other buff-colored insects crawled along the windshield as Old Rover carried them away from their grassland home. Soon they were joined by tiny denizens of the forest: grotesque horned spiders, green grasshoppers and bugs, bright and bizarre beetles, a huge green praying mantis, slow gaudy butterflies, and many others. As I appraised this collection, Old Rover mired down to the axles.

Simon revved the engine and tried to rock us forward. We sank deeper. We got out and left the doors open so the invertebrate riders could vacate the cab. We used an axe and my hunting knife to cut limbs from fallen trees to fill the ruts and provide traction for the worn tires. Then I pushed. And pushed. And pushed even harder. Old Rover finally lurched forward when the lower half of my jungle fatigues had run out of space for flying mud.

"The early British foresters of the Uganda Protectorate had this track cleared to survey the forest. No one but us have maintained it since Uganda became independent. The only traffic these days is this Land Rover and Tom's new one. The hell of it is I can't drive beside these ruts because of the stumps and logs hidden in the understory." Simon apologized.

Ten minutes later we were stuck again, this time on the Steep Hill. The Steep Hill was not all that big but it *was* steep and, due to the shade and the inch of rain yesterday (despite December being the start of the three-month dry season), the track was slick. We garnished the hill with more limbs, then roared up to the grassland.

"That's the worst place on the track," Simon admitted with

obvious relief. "Twice I had to spend the night here because I couldn't get up the bloody thing."

Meanwhile, blades of grass and seeds had stacked up against the radiator screen and blocked the flow of air. Water pumping back into the engine block was too hot to cool it. The heat gauge on the dashboard did not work because mice had eaten the wiring. As the engine heated, grass thrown on it from under the chassis started smouldering and the cab became sweltering. Most of the floor was already too hot even for a booted foot. A water temperature gauge would have been superfluous. We stopped again to clean the screen.

I caught myself staring at one of the bullet holes beneath the windshield. Green paint had peeled away where this bullet had punched through. During its younger days Old Rover had been used in the arid Karamoja District of northeastern Uganda to fight range wars and battles with cattle rustlers and elephant poachers. When Struhsaker bought this vehicle an official told him that during one skirmish Old Rover had been ambushed and a man sitting where I was now had been killed, probably by the bullet that punched this hole.

The grassland gave way to colonizing scrub, a thorny and nearly impenetrable association of fast-growing brush and small trees such as *Acanthus*. Given enough time and the right soil it might eventually become forest. Elephants and other large mammals sometimes favored it as a retreat from the sun and as a partial sanctuary from poachers. Kibale Forest Reserve is a mosaic of vegetation types. Tropical moist forest accounts for about 62 percent of its 215 square miles, but within and around the forest are islands of elephant grass and colonizing scrub, plus swamps in the lowlands. I ducked toward the center of the cab as spiny-leaved *Acanthus* whipped through the window. A red-wing francolin (partridge) scurried along a wheel track ahead as if to outrace us. At the last moment it took wing and veered into dense bush.

Again we emerged onto rolling hills covered with elephant grass. These grasslands were prevented from evolving to scrub, woodland, or forest by poachers putting them to the torch during the dry seasons. Finally, eight miles from the main road we veered west from the old forest track and crested the tallest hill in the neighborhood. Simon switched off the ignition. Steam growled viciously through the radiator cap. While we removed the radiator screen again to clean it, Simon glanced across the expanse of forest and mumbled, "superb."

We were atop a peninsula of grassland jutting westward into the forest. More islands of grassland emerged from the undulating ocean of tightly packed trees further to the west. Beyond that was the road between Fort Portal and Bigodi (a village near the Ngogo turnoff we had just taken). Twenty-one miles west of our hill rose the glacier-capped Ruwenzori, written of by Aristotle in 350 B.C. as the Silver Mountain which gave rise to the Nile, and by Claudius Ptolemy in A.D. 150 as, "the Mountain of the Moon, whose snows feed the lakes, sources of the Nile," but finally visited and revealed to the world by the intrepid Henry Morton Stanley and Emin Pasha many centuries later in 1889. Straddling the equator, this monolithic upthrust of the Great Rift System now tearing Africa asunder towers 12,500 feet higher than the 4,500-foot elevation of Ngogo.

The entire Kibale Forest Reserve slopes downward north to south. Kanyawara, 300 feet higher, was visible in the north. More than a mile toward the Ruwenzori, the camp at Ngogo was nestled on the small tip of a many-fingered peninsula of grassland jutting into the forest from the highest hills. Ngogo is the geographical center of the large new Kibale Nature Reserve (covering about 12 percent of the forest), perfect for a long-term study. Freeland's porters had cut an excellent trail-grid system running north to south and east to west, mostly at one-hundred-yard intervals, within two square miles of forest. This forest was virtually on the doorsteps of Ngogo's three tiny grass thatched mud huts, the only work of man visible in this primeval chunk of vanishing Africa.

The radiator had cooled, so we continued downslope, bouncing over tips of buried boulders and rolling over dead limbs of acacia studded with long thorns that would cause many flat tires in the future. Simon stepped on the brake pedal. It glided to the floor. He muttered, "Bloody hell! No brakes. . . ." This was one of Old Rover's protracted death throes. (Later I crawled under the chassis to discover that a rear brake line had a hole abraded in it.) The hand brake did not work because a leaky seal in the rear of the transmission had allowed oil to saturate the emergency brake

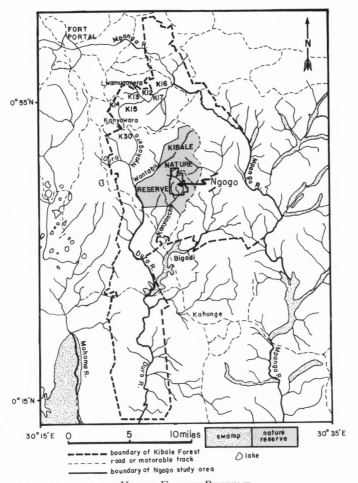

KIBALE FOREST RESERVE

shoes on the drive line. With the transmission in low gear, engine compression should have acted as a partial brake, but an overhaul was overdue and the compression was too low.

Simon looked at me with disgust in his eyes—possibly hoping that I had some trick up my sleeve but probably just to make sure I fully appreciated our current predicament—then he quickly turned to wrestle with the steering wheel. Old Rover squeaked, rattled, bounced, then roared down the hill so fast I thought the engine would tear itself apart. We were heading directly down a slope that ultimately lost at least 600 feet in height before leveling

23

out in a creek course, but, of course, some huge tree would stop us before we got that far. Fortunately the option of bailing out never looked inviting. And, even more fortunately, a saddle in the cross-cropped sward of camp slowed us enough to avoid plunging down the steeper section of hill into the forest.

The Ngogo crew had been anxious to meet me because in just a few months I would be running the camp, but they became even more anxious when the Land Rover roared down the hill without brakes. They walked over to the vehicle half expecting that our arrival had been one of Simon's little jokes. With their usual dignity and formality each introduced himself to me as we solemnly shook hands. Ben Alfred Otim and Medan Mukasa were game guards employed by the Uganda Game Department but paid through funds from Struhsaker's grant. Stephen Mugisa Yongili was the headman of camp and Clovis Okido was the porter, or trailcutter. None was from the same tribe as another; they constituted a mini-Ugandan national assembly. Greetings in Africa are not dispensed with lightly. Mukasa and Clovis did not speak English (Uganda's national language), giving me my first mandatory opportunity to use Swahili (which I had studied for a year), "*Habari gani* (what news)" answered by "*Mzuri sana* (very good)."

Considering that Simon had been here a year already, camp was fairly simple: three small mud huts and a grass latrine hut. During the brief period of light remaining I looked at the site where Simon's grass hut had stood. Phoenix-like, a new frame-work of poles had been erected by Yongili and Clovis to stand in the old hut's ashes. Part of an iron roof had been nailed down.

As I stood on the edge of the grassland to gaze across the vast expanse of trees, night fell. Luminescent beetles drifted in the darkness settling over the forest. Their glowing beacons switched on and off in species-specific patterns. Some of these advertised for a mate; others pretended to be mates so as to lure and devour the firefly who was fooled. Somehow they reminded me of one of the last things my grandfather had ever said to me before he died, "All that glitters is not gold." This maxim stayed with me as, surrounded by a symphony of unfamiliar night sounds, I made my way to one of the small huts and slipped into my sleeping bag. Familiar doubts about the future of my work delayed my sleep.

In the Forest

Sun dappled the forest floor. I was accompanying Simon in his search for the mangabeys of F.F. Group. We had turned so many times at the junctions of trails that my sense of direction was still one turn behind us.

We stopped beneath twenty-two large, charcoal-colored monkeys (twenty pounds for an adult male). They gave us only a brief glance and continued to scan for food, sit, eat, groom, climb lofty boughs, and pose in the distinctive stance of their species, a straight-legged posture with a dishevelled tail arched in profile over the back. It was almost as if the multimale group was unaware of us. Actually they were totally habituated to humans.

I glanced up to a mossy limb where a juvenile mangabey play-wrestled with a young redtail monkey. Redtails are about half as heavy as mangabeys, and, in contrast to the latter's drabness, they are white along their bellies, dark brown on their backs, and they have long tails of burnished copper, white cheeks bisected horizontally by a black stripe, and noses tipped with a white, heart-shaped spot. They are the most common primate at Ngogo. Because of the white spot on their noses, they reminded me of clowns. In fact, each new one I saw, and I eventually saw thousands, seemed more clownish than the last.

Sooty charcoal grappled with brown, white, and copper. In a kind of stylized waltz each monkey pulled his opponent toward him and mock bit him on the shoulders. After a few tug-of-war surges, the redtail leapt into the air to land approximately where the mangabey, who had instantly usurped the redtail's spot, had been a split second before. The redtail glanced at me in uncertainty, but his sooty playmate paid me no attention. They resumed their muscular waltz fifty feet above the forest floor.

"Have you ever seen that before?" I indicated the wrestlers.

Simon studied them for a moment then admitted, "No; that's a first."

The mangabey disappeared higher into the canopy; the redtail followed. Later I often saw groups of arboreal monkeys traveling in coordination with other species, sometimes due to a common attraction to a large fruit tree.

"Both redtails and mangabeys eat these large fruits of *Monodora*, but the redtail can't open an intact fruit," Simon explained as we watched the monkeys scramble through the trees. "So they wait until a mangabey bites a hole in one, feeds, then abandons it."

As Simon explained this I looked up into a *Monodora* and watched as a mangabey climbed away from a partially eaten, bowling ball–sized fruit. As if she had been listening to us, a redtail quickly arrived to insert her muzzle into the sticky pulp.

But because these associations often occurred without the attraction of a common food, they call for some additional explanation. Redtails, for instance, accepted any primate except male chimpanzees as a companion and frequently foraged with multimale groups of red colobus monkeys, the second most common primate at Ngogo. (Red colobus in Kibale are not really red but atop their brown and buff torsos are carrot-topped heads . . . heads that seemed to me too small to house even an average-sized primate brain.) Red colobus have bulging bellies adapted to digesting large quantities of leaves, a low energy but reliable food. Because redtails favor fruit, insects, and blossoms, competition with red colobus is slight.

Most likely the advantage gained by two species of monkeys remaining in proximity is explained by yet another perceptive sociobiological analysis by W. D. Hamilton and is explained in his essay, "Geometry for the Selfish Herd." The advantage works in two ways. First, an individual has only its own eyes with which to locate a predator in time to escape. So whenever it is foraging, grooming, or scouting a travel route it is vulnerable. But a group has many eyes, and the possibility that someone will be alert while others are busy increases everyone's chances of survival. The second advantage relies on the "you first" principle. Because predators often capture members on the edge of a group, many individuals tend to move inward in groups to surround themselves with others, as if to say, "you first."

Of course, the cost of more eyes for security and more bodies as a barrier is more stomachs needing food. But by associating with fifty edible red colobus, who are not particularly vigilant but do spot some predators, the redtails pay a small cost for security from powerful African crowned eagles soaring on broad wings immediately above the canopy and searching specifically for monkeys. Struhsaker told me each species of monkey reacts appropriately to the eagle alarm calls of its neighboring species, so it is likely that this form of mutual advantage is instinctual, having been reinforced by natural selection over untold genera-tions. But, as the redtail skulls I later found pierced by the outsized talons of African crowned eagles attested, the system is not foolproof.

Solitary monkeys probably live a touch-and-go existence. Struhsaker also discovered that in redtail society an adult male lives either alone or with a harem of up to ten or so females plus their offspring. From the perspective of the harem females, who Tom found to remain in their natal group and probably are related by matrilineal descent (by far the most common pattern among nonhuman primates), the harem male is little more than a convenient breeder who has proven his mettle, rather than lord and master. Because not enough harems (more accurately called matrilineal kin-groups of females) exist, most adult males are usually solitary—and vulnerable—until they can usurp a harem male. From the harem's point of view the "best" male is one who can successfully defend the harem from competing males the longest, because when a new male takes over he sometimes kills the youngest infants in the group.

This is another example of what sociobiology would conclude as the cold-blooded elimination of alien genes by male primates and it is explained, with some contention, in basically the same terms that I postulated in explaining why some male chim-panzees murder members of other communities. The reproduc-tive advantage gained by male redtails killing infants, however, is an immediate drastic change in the physiology of the mothers of the murdered infants. The females shift rapidly from lactational amenorrhea (which halts the estrous cycle while an infant is being nursed) to estrus. In other words, they become sexually receptive and ready to breed again (sometimes with the killer) almost immediately instead of several months to a year after the previous harem male was usurped (by which time the new male himself might be usurped). And despite the heroic defense these mothers put up against infanticidal males and the wounds they receive from them, they do mate with the killers. They are unfor-tunately trapped in a reproductive battle shaped by competition between males. Arguments against this sociobiological explana-tion basically claim that not enough incidents have been wit-nessed to rule out infanticide as an aberrant act by a few psycho-pathic males. (As I write this, however, such infanticide has been observed in at least a dozen species of nonhuman primates, including three in Kibale: redtails, blue monkeys, and red colobus in which a total of ten instances were observed involving six males.)

As we followed beneath the mangabeys as they scoured the canopy for food, Simon occasionally stopped, pointed to a tree, and told me its scientific name. Learning the seemingly endless variety of trees at Ngogo was essential for me to record the feeding habits of chimpanzees. But I felt slow at it. I had never studied plant taxonomy in my curriculum as a biological ecologist; I had always learned trees, like those in the Sierra Nevada where I grew up, by standing back and looking at them. But in the dense forest at Ngogo most trees consisted of trunks rising to intermesh with the crowns of other species. I could not see the trees for the forest.

So I had to learn them by their trunks. Trunks of different species were smooth, knobby, fluted, scaly, fissured, buttressed, perforated, patchy, flaky, straight, undulate, squat, elongate, thin, massive, brown, tan, gray, green, or nearly black. And if the trunk's surface did not provide enough clues, cutting away a small chunk of bark to reveal the slash often did. These were dry or they exuded a colored or clear latex. The slashes were striated, uniform, banded, grained, speckled, crimson, tan, white, green, or corky. Some looked like sliced bacon and some like clear pine, but all of them looked like something and many of them smelled like something. I started a new section in my journal describing the characteristics of the trunk, slash, foliage, and, if present, the blossoms and fruit of each new species.

For this, my first day at Ngogo, and the next, Simon and I divided our time between the forest and camp. The framework Yongili and Clovis had erected for a new hut was about twenty feet square. The first half of the roof had been nailed down with insufficient overlap between corrugated sheets, so it leaked like a sieve during the hundreds of tropical storms to come. While we nailed down the second half of the roof, this time with more overlap, Yongili and Clovis hauled water in plastic jerry cans from a tiny spring-fed creek a few minutes inside the forest to mix mud and pack the walls with it. While we worked, the rare wails, shrieks, and pant-hoots of chimpanzees occasionally eclipsed the other forest sounds reaching our hill. When they seemed close enough, I dropped my hammer and set off to follow them.

My searches soon developed into a pattern, but not a very fruitful one. I normally took a compass fix on the direction from which I reckoned the pant-hoots (the maximum long-range calls of chimpanzees as defined by Jane Goodall in 1968) were coming, then I entered the forest and tried to follow a route along the trail-grid system that took me the distance I had estimated. Several problems arose from this, but I was too green concerning chimpanzees to know which problems had to be taken into account and when. First, chimpanzees pant-hooted while traveling (I would learn why during the years ahead). So even if my estimate of the location of their source was correct, they often were gone by the time I arrived. Second, I was then underestimating how far pant-hoots carried; I was guessing not further than a mile. Consequently, much of my dead reckoning ended up short. Third, once in the forest, visibility was extremely limited by the dense profusion of vegetation competing for a place in the sun. A view of seventy-five feet through the tangled understory along the forest floor was exceptionally good, although visibility toward the canopy was better. Fourth, because I really had no idea where chimpanzees were likely to be, I missed checking important food trees and probably stumbled onto others so obtrusively that the apes sensed me and fled before I was aware of them. In short, I was getting nowhere. And that kind of progress was not encouraging.

After nearly a week of this I finally asked Simon his opinion. We had moved into the new mud hut and were sitting on his elephant grass bed with our nightly bowls of steaming mashed beans over mashed potatoes laced with limp cabbage. "Do you think it's possible to study chimpanzees here?"

Flickers of light from our feeble paraffin candle made his face shimmer as he looked at me, as if trying to figure out how to say what he was thinking. He pinched some coarse salt (our only spice) from a bowl and carefully ground it between thumb and forefinger on top of his dinner. "No, I don't," he admitted with a sigh. "They're even worse than the baboons. And at the rate *they* let me collect data, I figure it would have taken me forty years to do my dissertation research. Baboons and chimps are afraid of us

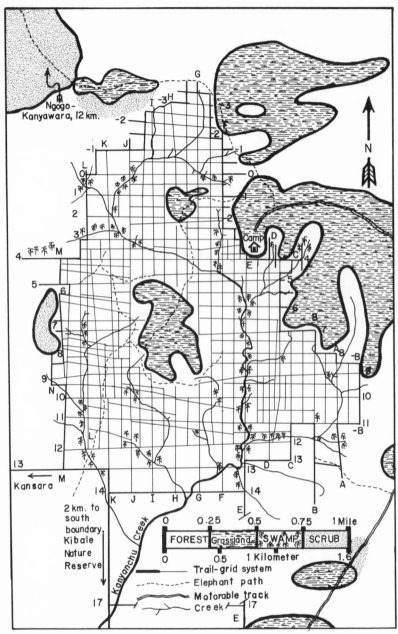

NGOGO STUDY AREA, KIBALE FOREST

and are too fast on the ground. There's no way to stay *with* them like you can with arboreal monkeys until they finally get used to you. So you can't habituate them. I think it is impossible." He paused for a moment, maybe to soften the blow, then added, "But if you manage it, you'll make a name for yourself in Zoology."

A couple of days earlier Tom and Lysa had arrived and pitched their big tent next to the rain gauge on the center of the close-cropped sward in preparation for one of their week-long sojourns at Ngogo. Their purpose was to record their respective observations on red colobus (Lysa) and redtails (Tom) simultaneously with Simon's observations of mangabeys during five-day stints they called "samples." A sample entailed following a group of monkeys for five days and systematically recording their behavior, foraging, and ranging in space at fifteen-minute intervals during the eleven-and-a-half hours of sunlight between consecutive sleeping trees. In the capacity of my local supervisor, Tom firmly suggested that I accompany Simon during the impeding sample to learn to recognize Simon's (aka Bill Freeland's) mangabeys individually and to become familiar with the sampling techniques.

Even as early as in Kanyawara Tom had advised me that he considered it best that I "assist" Simon during part of the next five months. Because all Simon really needed were data on ranging, which during five or ten days each month he collected quite well on his own, and because Simon's biggest problem was finding enough novels to read to kill his remaining time, I was dubious about the need for my assisting him. Simon and I concluded that Tom was patently preparing me to take over the mangabeys when Simon left off; Tom even mentioned that possibility to me. It was beginning to seem that, despite his letter to the contrary, Tom shared Simon's (and the rest of primatology's) dubious appraisal of my chances of studying the apes of Ngogo. Why then did he encourage me to come to Kibale?

At first this apparent contradiction made no sense to me. But soon I saw that Tom's continuing research projects in Kibale required that some scientist be constantly present, preferably one each at Kanyawara and Ngogo, to act as supervisor to maintain the study areas, insure logistics for the crews, and direct anti-poaching patrols. After Simon left, only Tom and I would remain, but because Tom and Lysa regularly spent three or four months each year outside Uganda, I was an asset regardless of what I managed with chimps. Eventually I came to understand that in his letter to me about being able to study the chimps, Tom was thinking of the chimpanzees at Kanyawara, who were in fact nearly completely habituated to people, but who also ranged through a mosaic of severely modified habitats created by the felling of timber and official poisoning of "weed" tree species with arboricides. Their unnatural ecological situation made the Kanyawara chimps unsuitable as a primary focus of my proposed research. That was why I was at Ngogo. But the upshot of the system in which I found myself was this: I was an asset; a chimpanzee project would be icing on the cake. Again I felt a nagging doubt that the insiders knew something I did not.

Mangabeys are interesting, mildly, but each time I heard a distant pant-hoot I could barely control my urge to chase it. Before dawn on the fourth day of the sample, Simon and I were gazing up at scattered, indistinct blobs perched in two tall trees and waiting for them to coalesce into mangabeys. Pant-hoots split the somber morning and I set off again in hope of first contact.

Even after more than a week I knew only a fraction of the forty-five miles of trails and even less of what lay between them. I almost tiptoed in my stealth, zig-zagging back and forth across the trail-grid system and searching the trees. No new calls gave me a fresh clue. It took me two hours to cover half a mile.

Something thudded against the ground—fragments of *Monodora* fruit. I edged closer. Silently, the wrinkled and white-bearded face of an old male peered at me from an open space in the crown. Tight-lipped and inscrutable, he studied me for about three seconds. My heartbeat quickened. I knew it was impossible, but I desperately wanted to communicate my friendly intentions and the merits of my mission, "I come in peace. . . ." It felt exactly what I imagine it would be like to meet an intelligent alien. But this chimpanzee was not a naive alien. He took no chances. Surprisingly silent he quickly climbed through the crown and swung to the ground. Then he was gone, swallowed by the forest.

I was stunned for a moment. This was not like meeting any other wild creature for the first time, no matter how rare, dangerous, or beautiful. Chimpanzees are so manlike and their intelligence so superior to that of other mammals that the initial encounter is much like an eerie glimpse through the time-misted veils of our own evolutionary past. Evidently, though, the fascination was not mutual.

I walked under the *Monodora*. Fragments of thick-hulled fruit were strewn beneath it. A fresh pile of chimp dung nearby had already attracted dung beetles. *Monodora* seeds were scattered through the dung like peanuts in chocolate. The pulp of the fruit is so sticky that it is easier to swallow the seeds than to spit them out. In many cases, after passing through the alimentary canal seeds are still viable, perhaps *more* viable than if they had not been scarified by passing through the gut. During the apes' wanderings seeds often are deposited far from their source. The tree hardly could ask for better cooperation for dissemination, but they do have to pay for it by producing nourishing fruit.

That chimpanzee renewed my determination. From that day on I quit Simon's samples and continued to plod through the forest toward distant pant-hoots and screams like the proverbial donkey stepping toward the dangling carrot. I soon realized that, instead of knowing where they are when they pant-hoot, what I really needed to learn was where they wanted to be most. Food had to be the answer. Every published report of chimpanzees agreed that fruit was their favorite food. I started mapping the locations of fig trees at Ngogo and checking them frequently for the presence and ripeness of fruit.

Meanwhile, many days passed without my even glimpsing another ape. Other primates seemed to be everywhere. In addition to red colobus, redtails, and mangabeys, I had seen black and white colobus, blue monkeys, baboons, and the mysterious terrestrial l'Hoest's monkey about which nearly nothing was known. I started a series of monthly censuses along set routes within the trail-grid system. For each sighting of primates or other large mammals I recorded their location, behavior, numbers, and whether they were in association with other species. I eventually determined that monkeys outnumbered chimpanzees 197:1. No wonder they were hard to find!

My days blended into a succession of wanderings along trail after trail. I found most fruit trees empty, and even distant pant-hoots were rare. The complexity of this shadow world beneath the canopy of the forest still baffled me. Finally I found the apes again, five seconds before one of the party of three spotted me. He immediately stopped eating *Monodora* fruit and silently slid down the bole of the tree to vanish. Suspicious of the sudden defector, the other two peered around until they spotted me, then wasted no time in following their companion.

For five days in a row I contacted them like this. I even tried concealing myself during contact, a modification of my original plan to remain visible at a respectful distance. But hiding was not the answer because I could not really see what the apes were doing, and because I would never gain their trust that way. Without trust and tolerance I would learn little about them.

I needed a means of approaching the apes or exposing myself to them wherein I would appear so harmless that they would remain long enough to learn that I actually was harmless. This required a place where food was available in such abundance that they would be very reluctant to leave. It was easier now to understand why other researchers had resorted to setting up feeding stations to win them over.

Why did the apes fear me as such a deadly threat? It was not me personally they feared, but me generally. People sometimes hunt chimps for food in most forests where they occur, but hunting was prohibited in Kibale. This prohibition, though, was only as real as the enforcement, which unfortunately was nearly nonexistent. Poachers in Kibale did not even bother to wink at the law when going after buffalo and elephant with their illegal firearms. The last hippo wallowing in the Dura River, a stream draining Ngogo, had been shot only a few years prior to my arrival. The few surviving elephants and buffalo now stampeded at the hint of a human.

I learned that while these were being decimated with guns, gangs of up to forty small-time poachers in some parts of the forest and along its edge hunted with nets stretched for hundreds of yards. Men with dogs drove bushbuck, red duikers (tiny antelope), blue duikers, bushpigs, giant forest hogs, young waterbuck and buffalo, baboons, and occasionally even chimps and l'Hoest's monkeys into the clinging mesh. Other men posted there then rushed forward to plunge the iron heads of their spears into the vitals of their prey.

Yongili, our twenty-year-old headman from Bamba (the district at the northwest foot of the Mountains of the Moon), became a source of recent history for me. He explained that he had been told that, prior to the intertribal war that raged across Toro District in 1962, the Bakonjo (who inhabit the eastern slope of the Ruwenzori) and the Bamba sometimes killed chimpanzees and other primates in Kibale Forest and considered them toothsome fare. After the Bakonjo and Bamba were vanquished by the Batoro, the former tribes retreated to their tribal territories hugging the Ruwenzori foothills. Quite unlike their enemies, the Batoro regard primates as unfit for human consumption and may even refuse to use the cup, bowl, or cooking pot of a person known to eat them. After 1962, the bottom had dropped out of the primate meat market, so chimpanzees and monkeys were no longer hunted purposefully. But the indiscriminate process of being driven through the forest toward the net and waiting spearmen continued.

A chimpanzee can live forty years or more. Some of the older adults at Ngogo could have witnessed relatives or companions dying under the spears, bullets, or poisoned arrows of the Bakonjo. Many apes must have experienced at least part of the terrifying net drives because the Batoro poachers normally let them escape, perhaps repeatedly. It is small wonder that the apes of Ngogo fled from me as if I were an enemy. But to test the hypotheses explaining killer apes and to conduct the ecological research necessary to make sense of their social structure at all I *had* to observe them. The question uppermost in my mind was, would they continue to flee indefinitely? Were Simon and Tom and everyone else correct in their assessment? Or could I win their acceptance?

— 3 —

Stalking the Wild Fig

R ain had soaked the forest. I stepped between the mist-shrouded trees at dawn and headed down the slippery path. Each time I jostled the foliage I precipitated a sudden chilly shower. But I was too preoccupied to pay much attention: if ever I would gain the tolerance of the apes of Ngogo, my opportunity was now.

Yesterday, while returning from the southern boundary of the trail-grid system in the subdued light of dusk, I had passed a huge fig tree at the base of the steep hill only a third of a mile from camp. I had been checking this tree daily because it was loaded with green figs. *Ficus mucuso* was its scientific name. It is a rare species, but one of the very largest forest trees in all of Uganda. And this particular tree was the largest of the more than one hundred I found during my years in Africa. Its massive bole was supported by flaring buttresses beginning nearly twenty feet up the trunk and then wandering another twenty to the sides. It looked like a rusty rocket ship with warped wavy fins standing forgotten in the forest. Its bole shot up about sixty-five feet then spread its gigantic limbs hydra-like to produce a hemispherical crown as wide as half a football field. I conservatively guessed that tens of thousands of figs larger than apricots were strewn in clusters in this auditorium-sized crown. In my survey for potentially important feeding spots for chimpanzees I had never seen anything like this, but day after day these figs had been green. But at dusk yesterday, exactly one month after my arrival at Ngogo, chimpanzees were finally picking them.

One female spotted me after twenty minutes and gave a low, clear "hoo" or "hum," the chimp equivalent of a human "oh" uttered in surprised uncertainty, or when something unpleasant is about to happen. Then she descended to the ground. The others took their cue from her and peered in my direction. In my standard effort to appear as inoffensive as possible, I scratched myself vigorously, which chimpanzees often do when uncertain and unaggressive or simply bored, and I was careful not to threaten them by staring up in the tree; I glanced up frequently but spent most of the time examining where I was scratching and looking around me. Despite my reasonably good manners, they chorused a pant-shriek at me and two of the females threatened me by vigorously shaking branches. Then they hastily climbed through the crown to slide fireman-style down a slender-boled access tree.

Twilight revealed one adult female alone in the *mucuso*. From 100 feet above the forest floor she kept a wary eye on me, alternately stuffing figs into her mouth and hooing. I continued my scratching and wished that I had a real itch. Eventually she quieted down and gave the figs more attention than me. Finally she turned her back on me and just ate.

I had a hard time trying to decide whether this indicated trust, indifference, or hunger, but her priorities were clear: food was more important than my presence. I was able to observe her, or I would have been had it not been nearly dark. This tree was the place I needed. I could hardly wait for morning.

Now, during the final stage of my approach to the tree this morning, I chose the side of the creek opposite the *mucuso*. The last obstacle was a smooth fallen tree six feet thick and slick from rain and slime. It was overgrown with a tangle of vines and dripping foliage that waved like a giant flag when I touched it. I paused to study it. I did not want to be seen yet. I was afraid of commiting some error, a simian solecism, that would cause the chimpanzees to permanently abandon the *mucuso*. Would that female now faintly visible fail to detect me?

Her back was toward me as she munched figs. I inched my way up the log, then lost traction and slid back down to thud against the soggy detritus. Another personal rainstorm added insult to injury. Up again. From the log I scanned the *mucuso*. She gave me a hasty last look before being swallowed by the vegetation.

I jumped to the ground and walked closer. To the right of the trail were two shrubby trees surrounded by African ginger, a common understory herb that looked like a thin-leaved iris and produced pungent leaves I often crushed for their aroma. African ginger also produced a small fruit that baboons and chimps ate. I slipped between these two trees to find partial concealment from the fig tree 200 feet away. It was a good spot. I made a nest by bending some African ginger and ferns on top of one another, then I squatted down and stared at the tree.

An hour and a half later a dark, hairy arm suddenly appeared high on a slender tree leading into the *mucuso*. An adult female paused, gazed around the crown, then climbed into it to forage. A juvenile female followed but stopped to feed well apart from her companion. They both foraged as if they had not eaten in days and as if oblivious of one another. Neither glanced in my direction. I sat with my eyes glued to my field glasses and felt like a spy. A happy spy.

I was on the west side of the fig tree. The sun rose above the eastern hills to shine through the *mucuso* into my eyes. I was afraid sunlight would reflect off my lenses and reveal me, but it was too late to move. Besides, this was the best vantage point. I was using a pair of Leitz Trinovid 10 × 40 field glasses, not only the most expensive piece of equipment I owned but absolutely the most essential. The lenses were so perfect I could watch through them for hours without eyestrain. But, as I sneaked another quick look through them at the adult female, she suddenly stared intently in my direction, then clambered directly to an exit route and quickly slid down. The juvenile watched her retreat, then followed.

I sat back on my nest and continued to gaze toward the *mucuso*. I was disappointed that she had spotted me so soon but eventually she would have anyway. My plan *depended* on the apes spotting me. I wanted to observe as many of them as possible before they noticed me, of course, but then bank on them returning repeatedly, finding me again, then eventually electing to feed despite my presence. I studied the *mucuso*; it was by far the largest concentration of fruit I had seen in the forest. Without this incentive to stay I would not have a chance. And if this did not work, I did not know what would.

Noon came and went. Then I waited for another six hours in anticipation of the dinner rush. I speculated on the future. Meanwhile a hard rain spattered against my U.S. Army poncho and found pinholes I did not know existed. No more chimps arrived. I tried to convince myself that it was because the figs were still too green.

While hurrying down the path next morning to return to my vigil at the fig tree, I wondered whether again only a few apes would visit the tree, whose green figs now were definitely turning to yellow then orange (full ripeness was nearly red). I was unprepared for the crowd that did appear.

Four females were in the tree when I arrived, but they saw me almost immediately and fled. Another crummy start. But this day turned out differently. For the next ten hours at least ten parties containing one to nine chimpanzees visited the *mucuso*. I was hard put just to figure out how many were up there and which age, sex, and reproductive class best described them. They arrived in maternal groups of females plus their young, mixed groups of males and females plus young ones, and in all-male parties.

Their comings and goings confused me. Keeping track of the different individuals, even considering only when they arrived and departed, was a chore. And despite my field glasses, I was not sure my identifications would be reliable for recognizing them in the future. Sometimes sixteen apes moved through the crown at once and left me in confusion. It was the happiest problem I had had in a month.

But even this early during my first *mucuso* vigil some patterns already were apparent. Sometimes when adult males arrived together, they joined in earsplitting choruses of pant-hoots, shrieks, and wails, often drumming on the tree's buttresses before climb-

ing up. Depending on the acoustics of the terrain and the elevation of the listener, displays like these could be heard for more
than a mile. As with nearly everything else chimpanzees have
been seen to do, this behavior has been interpreted differently by
different people. One explanation for these calls during arrivals
at fruit trees is that they announce the existence of the food to
neighboring apes. This implies something akin to altruism, especially if one believes the apes cannot be certain who will hear
them and then arrive to compete for food. Another explanation is
that choruses of pant-hoots warn alien chimpanzees of the presence and numbers of their competitors, and are given at fruit
trees because this is one of the times when several males are
together and consequently sound more impressive. In 1971 Jane
Goodall wrote that pant-hoots seemed to act as a contact call. At
this point the several functions of pant-hooting were unclear. But
I was certain of one thing: they were far louder than any noises I
could produce with my own lungs.

But my interest spurred me to start a written log of every
episode of pant-hooting I heard during twenty-four hours of each
day, noting the number of apes present, the context, which
chimp initiated the vocalization, and which ones chimed in.
(When I was not actually observing the authors, I estimated their
minimum number and location.) Eventually my analysis revealed
that pant-hoots were more than by-products of exciting situations. First of all, males initiated pant-hoots ten times more often
than females, and males responded to the distant pant-hooting of
other chimpanzees three times more frequently. Relatively, females initiated very little pant-hooting. But pant-hooting was
contagious. When an adult male started his initial buildup by
uttering a long series of rising eerie wails and clear hoots, other
apes commonly joined in almost as if they could not stop themselves. In the ensuing cacophony the initiating male gave clear
rhythmic hoots, a second gave piercing rhythmic screams, while a
third wailed, barked, or otherwise improvised. Sometimes as
many as eight or more apes joined to produce one of the auditory
experiences of a lifetime for a human within point-blank range. It
was my distinct impression that each ape contributed a sound
slightly different from anyone else's, as if to create a chorus
which would allow a listener to count heads by ear.

Why do chimps pant-hoot? More than half of nearly 1,000 pant-hoots I recorded at Ngogo either elicited additional pant-hoots by a separate party some distance away within three minutes, or they were elicited by such hooting from a separate party. In short, separated parties of chimpanzees were communicating with one another across the forest. What could they have been communicating?

Peter Marler and Linda Hobbet recorded pant-hoots at Gombe and analyzed them sonographically to discover that those of adult males differed significantly from those of adult females or juveniles. They went on to conclude that several elements within the four-phased series of wails, hoots, shrieks, and roars indicated that pant-hoots are recognizable *individually* both to humans and to other chimpanzees. Eventually I began recognizing the pant-hoots of some of the apes of Ngogo. With their superior ears and experience chimps must be capable of at least similar discernment. And, as I mentioned above, even if one cannot distinguish exactly who is pant-hooting, one can often estimate a minimum number of apes involved. So, in answer to the question of what chimpanzees communicate by pant-hooting, it seems fairly certain that pant-hoots convey the following: the identities of some of the callers, the number in the chorus, the location of the party, and, if repeated while on the move, the direction of travel. The energy devoted to pant-hooting suggests it is an important communication. So now the question shifts to: why do the apes expend so much effort in communicating with one another?

As might be guessed, the answer is complex, but the question itself contains another question buried within it: why are chimpanzees of a community separated so often? And the answer to this one is based in ecology, which is discussed in detail in the next chapter and others. A condensed explanation, however, would be this: chimps prefer foods, like fruit, which are rare and scattered in food patches that normally contain too little food for the entire community. Consequently the apes maintain constant relations with only a few others and forage in small parties or alone, especially during lean times (44 percent of all the parties I saw at Ngogo during my research consisted of only one chimpanzee). But despite the limitations of food dispersal, chimpanzees obviously prefer one another's company, and when they can afford to travel together they do. The complexities of why this is the case is the main thrust of this book.

When traveling in a large party exacts too great a metabolic toll because of food dispersion, separated parties can keep in touch by pant-hooting, which announces the option to join distant callers. And while it is not possible to say that pant-hooting serves this function first, it definitely keeps parties in touch. Even so, many times I observed chimps who listened to enthusiastic pant-hooting choruses wafting across the forest but then remained silent and turned down the option. Of course pant-hooting probably also functions to warn away alien chimpanzees and thereby avoid or prevent a dangerous physical confrontation. Pant-hooting serves several functions varying from situation to situation, but the most amazing one I determined was food calling.

The uproars created by males visiting the huge *mucuso* I was monitoring announced across a wide expanse of forest that something was happening. Other chimpanzees could not help but hear, but why were they supposed to hear? Or, in other words, what were these males trying to say? It seemed to me that more subsequent parties appeared after these vocalizations than if none was given. I could not collect enough data from this one *mucuso* vigil to answer this question, but over the whole of my research I did collect enough.

I compared the frequency of subsequent arrivals after pant-hooting males had arrived with the frequency after the arrivals of quiet males. The upshot is these bedlams of pant-hoots and buttress-drummings *did* alert other chimpanzees, who then were significantly more likely to arrive at the same food tree than if no calls had occurred. These were food calls. Their very nature implies a complex society unique among nonhuman primates.

What do males gain by inviting competitors? If their behavior is pure altruism and they gain nothing, we would expect natural selection to have eliminated loud male chimpanzees near food sources long ago. Also, at first glance, such altruism by males seems somewhat at odds with the concept of their being killer apes. But sociobiological analysis blasts the possibility of true altruism. Males who gave food calls were selfishly inviting other apes who potentially benefited the callers in three major ways: (1) Depending on who showed up, a caller benefited through social interactions such as mutual grooming, or gained in security by subsequently traveling in a larger party. (2) If the arrivers were genetically related to the callers (likely when it was another male, assuming male retention among Kibale chimps), its better health due to improved nutrition and its consequently better reproductive success due to better health benefited the callers through inclusive fitness in ways I explained in Chapter 1. (3) When the arriver was a female in estrus, the callers had opportunities to mate which otherwise they might not have had. These advantages to food calling make it less surprising that females, who share fewer genes with the rest of the community, never give food calls or invite competition. They would gain less by doing so and, due to their lower dominance status, would be left with less to eat. But are these benefits to males really worth the increased competition for food? Or, more to the point, do males incur a real cost by food calling?

Not surprisingly, males arriving at relatively small fruit trees rarely pant-hooted; competition there would be severe. But in large fruit trees competition between the apes seemed minor. In short, observations of food calling clearly reveal that males are intelligent enough to routinely measure available food against the potential benefits of food calling and then make a selfish decision. Quite independently of my observations (and I of his), as a graduate student of Cambridge University working under Jane Goodall, Richard Wrangham (1975) observed males at Gombe giving food calls at large food trees.

It was a stroke of good fortune that this *mucuso* turned out to be the largest fruit tree I ever found. It was now attracting chimpanzees for repeated visits despite my presence, and by doing so provided my first breakthrough in habituating them. With one in particular, Raw Patch, I began a relationship not only valuable professionally but also personally fulfilling. When I first saw him, Raw Patch (R.P.) was traveling with three others who hooted, shrieked, and drummed the buttresses of the *mucuso* before climbing into it. He led the route up the egress tree but, as the others raced past him, paused at the threshold where it intersected the crown to gaze across the forest. Perhaps he was cautious; perhaps he simply liked the view. A thinking man's ape, I thought to myself. Then his scanning focused on me, an alien hominid squatting on a muddy pile of African ginger and holding strange black, exaggerated eyes in front of his face. R.P.'s eyes grew round and he dropped like a flash to gracefully plummet down the egress tree, seeming to barely touch one branch before tapping the next in a descent of eighty feet. None of the other seven chimpanzees remaining in the *mucuso* noticed me, and in their orgiastic gobbling of figs, they barely glanced at R.P.

He returned three hours later with eight companions and a bedlam of vocalizing. This time I had a bit longer to study him. His permanent canines had not erupted, an indication that he was two years from full adulthood (Jane Goodall's [1983] analysis of maturation at Gombe placed him at thirteen years old). So it was unlikely that he had much status among the prime males of the Ngogo Community. And unlike many of the Kibale chimps, who are included with the eastern, long-haired subspecies of chimpanzee, R.P. had very short, light colored hair, especially on his head, which had a sculpted appearance emphasizing his musculature. His name derived from a pink wound half the size of a hand on his lower back. He also had two other gashes across the center of his back. I wondered what kind of trouble had given him those wounds. Too soon, he and his companions spotted me and rushed out of the *mucuso*.

Simian foragers visited the tree more and more frequently, and by noon of Day Four R.P. arrived again at the top of the egress route to peer at me furtively from behind the foliage. Then he raced into the *mucuso*, grabbed several ripe figs on the run, crammed his mouth to capacity (seven or eight), and hurled himself down the exit route after less than a minute in the tree. His three companions followed him.

An hour later he and the others returned and repeated the entire performance. Another hour passed and they did it again. They made their fourth and fifth visits fifteen minutes apart and stayed only briefly. Obviously I was causing them inordinate travel fatigue; had I not been there they surely would have prolonged their stay by an hour or more. These lightning fig raids resulted from the conflict between wanting to feast on that bounty of figs and wanting nothing to do with me. As I hoped, and as their returning for these sorties demonstrated, the lure of figs was driving them beyond their fear of me. But I wondered how long their crazy pattern of visits would last before they tired of it.

Surprisingly, R.P. broke the pattern late that same afternoon. He arrived shortly behind two females, both of whom were eyeing me as if I were the worst sort of company. Before entering the fig tree, R.P. paused to gaze at me for a full five minutes, a lengthy appraisal by the standards of both apes and humans. He scratched himself profusely and was patently undecided. Then he abruptly climbed into the *mucuso* to forage for an hour. He concealed himself behind foliage most of the time, but I saw him occasionally. He had the entire tree to himself; neither female had been able to summon his bravado. R.P. glanced at me frequently and consumed figs as if he were in a contest. I scratched myself and acted disinterested.

Eventually he descended. His stay encouraged me that my project had a chance. In fact, R.P. returned on three more of the ten days I spent on vigil at that *mucuso*, and twice he stayed nearly an hour. And, although I did not see him again for almost two months, when we next met he had not forgotten me. This giant fig tree became my letter of introduction to several other apes. Stump was among the oldest of the forty-six chimpanzees I would recognize at Ngogo. Whereas R.P. was volatile and abrupt, Stump was slow and sure. He had to be. Not only was he probably at least forty years old, his left hand was amputated at the wrist, most likely by the heavy braided wire of a poacher's snare. When Stump first saw me he dropped into a crouch and shrieked, growled, and barked at me for almost half a minute. Then he carefully climbed out of the tree. But for some reason I never understood, he returned and quickly became blasé about me.

Stump traveled most frequently with a prime male whom I named Silverback (S.B.) for the light coloration of his lower back. (Stump traveled next most frequently with R.P.) S.B. was obviously of high status, though not of great size. And he found me intolerable. Almost invariably he abandoned food trees prematurely when he saw me. Exceptions occurred when two or more other males were present and acting nonchalant by ignoring me; most often these two others were R.P. and Stump. In such circumstances S.B. ignored me too. For fifteen months he avoided me assiduously, then one day when he was alone something snapped and he reversed course by ignoring me. I still do not understand it.

These observations of divergent behaviors by individual apes who often associated with one another pounded home the lesson that they were individuals, not cookie-cut primates with "hard-wired" responses to standard situations. Therein lies one of the causes and effects of the complexity of chimpanzee sociality. Chimpanzees clearly thought situations out, then acted on their own decisions based on personalities and past experiences. This individualism could have far-reaching implications. Soon I would discover that even the *communities* at Ngogo and Kanyawara exhibited social differences. Such variation between apes of the same *and* different communities forced me to wonder just what "standard" chimpanzee life is—and to reevaluate how difficult it would be to unravel the questions about chimps as killer apes. At this point, though, I still had to convince most of them simply to accept me.

At first it puzzled me that the most common chimpanzees to tolerate me were solitary estrous females. Females experience an approximate lunar cycle, much like that of humans. During ovulation and sexual receptivity (estrus), however, their external genitalia become engorged with fluid and are strikingly swollen. Doubtless this swelling is a visual sign that the female is ready to mate. But it seemed to me that along with these physiological changes, psychological changes also must have been occurring. Estrous females often calmly accepted me when other chimpanzees fled. I suspected that a psychological change would help a female reduce her normal caution and reluctance in being close to the more dominant males with whom she must breed. I suspect that this lowered caution carried over to me in a general way, one for which I was grateful.

One such female, normally solitary, repeatedly visited the *mucuso*. I named her BULB for bare-under-left-breast. She spent more than twelve hours in the fig tree during the final three days of my vigil and rarely was upset at my clumsy tests of her tolerance. Because she was so accepting, I looked forward to observing her often, but after the fruiting period of this *mucuso* ended I never saw her again. In view of my many subsequent encounters with the other apes of Ngogo, I concluded BULB was probably a transient from another community.

By contrast, mothers carrying infants became the least habituated to me. During two years at Ngogo only two mothers tolerated me; the remaining half dozen fled unless several adult males were nearby. But at Kanyawara mothers were much more tolerant. I expected juveniles at Ngogo to be curious, but they were extremely conservative and rarely wasted any time increasing the distance between us.

Slowly I became familiar with more individuals, and even more slowly some of them began to tolerate me. After more than one hundred hours of staring at that *mucuso* from my muddy niche I had observed 116 aggregations of chimpanzees averaging 3.7 apes each for a total of more than 30 hours. These statistics reassured me, even though I had seen little more than who was traveling with whom, protocols of foraging, and how they pant-hooted. I knew that this tree was just a beginning; I also was acutely aware that no further progress would occur until I found another place like this *mucuso*.

In fact, I had to multiply this vigil by many others. The giant limbs of this *mucuso* were now denuded of their figs. Butterflies and bushpigs swarmed over the discards heaped below. A new tree was critical. But to find it I had to learn to view the forest somewhat as a chimpanzee does.

From a strictly scientific perspective I had to monitor the availability of fruit at Ngogo to determine how it influenced the chimpanzees' interactions and patterns of travel. For the apes themselves the location of fruit is vital information. By now I knew many species of trees, but I was most impressed with Ngogo's fourteen species of *Ficus*. I had a hunch that figs were their favorites. My hunch was vindicated later by an analytical natural history article by Daniel H. Janzen titled, "How to be a fig." Most critical to the chimps here, Janzen pointed out that, in contrast to most fruits, which are high in sugars but too low in proteins to provide the requirements of primates, figs are high in proteins. The reason for this is figs contain insects, tiny symbiotic wasps which depend on a single species of *Ficus* for fruit in which to lay their eggs. Botanist J. T. Wiebes confirmed that the fig depends equally on the single species of wasp to pollinate the tiny flowers inside the fruit. In this symbiosis many wasp larvae end up inside the figs when they ripen and, inadvertently, provide the additional protein essential to the apes, other primates, fruit-eating bats, and many large birds, all of whom further the fig tree's greater goal of setting its seed throughout the forest.

So I spent the next week plotting a maze-like itinerary from one important food tree to another. Soon I was visiting more than one hundred trees at least once each month to record their phenology, or seasonal "behavior": their periodicity of fruiting, flowering, and leaf production. To determine whether and how much production was occurring I had to scan each crown entirely. Crowns usually were high above the forest floor and intermingled with and obscured by neighboring trees. Scanning them necessitated wandering through the understory and staring straight up into the back-lighted foliage through field glasses. This was such a pain in the neck that I soon grew to detest phenology. But the data were essential for me to begin viewing the forest through the eyes of an ape.

Eventually my phenology data revealed that the periodicity of the forest was irregular and that most of the figs were sporadic and unpredictable. Some other species were synchronous but again unpredictable. Kibale Forest had three months of rain beginning in March followed by three months of drier weather, then three more of rain, and so on. Despite the regular seasons, some species often did not bear fruit for years on end.

In short, fruit was patchy. Patchy is ecological jargon for a resource distributed in discrete locations in space and time rather than evenly. Patchiness is relative: fruit is more patchy than insects, which are more patchy than palatable leaves, and so on. Also, what is patchy to a rodent may seem uniform to an elephant. A species of forager which specializes on patchy and ephemeral resources is one of the rarest in nature. Fruit is both patchy and ephemeral but comprised the majority by far of the chimpanzees' diet at Ngogo (and in most other habitats in which they have been studied), which makes them specialists of a rare order. Lean times, when little fruit was available, were the greatest test of the apes' social system.

Success in the scramble for fruit demanded that chimpanzees do more than rely on bumping into it randomly. A well-developed spatial sense is essential for remembering the locations of food trees scattered among the tens of thousands of nonfood trees within the ten square miles of their home range. It is not surprising that captive chimpanzees, even as young as juveniles, tested by psychologist Emil Menzel, Jr. (1973), have demonstrated an unusual ability to choose, by memory alone, efficient itineraries to collect scattered, hidden caches of food. Even more intriguing, captive chimps shown by Menzel (1979) where a food item was hidden but then restrained from getting it on their own could somehow communicate its location to another ape so he could get it.

Although knowing the whereabouts of a fruit tree is absolutely vital, it is only half the information a chimpanzee needs if he ever is to harvest the fruit. Predicting *when* fruit will be ready is the other half. It is revealing that Richard Wrangham (1975) admitted that Gombe chimps were "good botanists" who searched in the correct plant communities for species in season. In the lab, with food as an incentive, chimps have even learned the value of money: how to earn it and how to save up to thirty tokens in anticipation of purchasing high priced foods. The incentive is both natural and imperative, and the apes' great intelligence is geared in part to excel in attaining it (see also Chapter 15).

Consequently the most important by-product of monitoring all my phenology trees and mapping every fig tree was learning the places where chimpanzees might congregate in the future. Nine days after quitting the *mucuso* I found a medium-sized *Ficus dawei* with ripe figs. The only regular visitors were members of a large family consisting of a prim old matriarch I named Farkle plus her three daughters, Felony, Fern, and Fanny, ranging in age from at least nine to less than one year. Sometimes they were accompanied by a subadult male, whom I assumed from their closeness was Farkle's son and whom I named Fearless.

For reasons I never understood, few adult males visited this fig tree during my vigils. I wondered if perhaps they were frequenting some other part of the forest too distant to justify the energetic expense of traveling here. The figs of this *dawei* soon were consumed by the Farkle family, a few other apes, and scores of mangabeys, redtails, blue monkeys, and baboons, so once again I had to search for new food sources for the apes. Now, though I had barely begun to scratch the surface of the lives of the chimps of Ngogo, I knew I was on the right track. But just when I started to shake my constant doubts about being able to observe the chimpanzees exactly as they were in the wild, Uganda suddenly exploded into trouble that threatened to end my research and possibly my life.

— 4 —

You Are What You Eat

"There were dead people along the road, so my friends gave up. . . ." This was part of the news that Otim's wife, Tedi, brought him from Bigodi, the small village on the fringe of the forest nearly nine miles from Ngogo. The villagers had begun the long walk from Bigodi to Fort Portal in hopes of catching transport along the way, but they turned back when they found the bodies of murdered people dumped along the sides of the road. Although they were not identifiable as such, the victims were Acholi and Langi, the so-called warrior tribes of Uganda. Because Otim is a Langi, Tedi feared for his life and found a man to accompany her on the hike from her uncle's home in Bigodi (a hike most Ugandan women would shun because it penetrates the jungle) to warn Otim he was marked for death because of his tribal heritage. I wondered how Uganda had reverted to the savagery of the nineteenth century.

Idi Amin Dada, the current ruler of Uganda, was born a Kakwa (a small West Nile tribe) in Ugandan barracks of the King's African Rifles (KAR), a colonial force manned by British officers commanding native troops. He enlisted as soon as he grew large enough to justify getting on its payroll. During Kenya's Mau Mau uprising he was a zealous soldier who quickly gained a reputation for brutality that augured his future. Amin also became the champion boxer of East Africa. He rose through enlisted ranks to earn a commission in the KAR, which later placed him in the position to order the penises of recalcitrant Karamajong rebels cut off as a lesson in authority. By the time Uganda became independent in 1962, Amin's sixteen years of service had escalated him to the leading ranks of the KAR.

Years of political machinations by Dr. Milton Obote, Uganda's first elected prime minister, to consolidate power from an imbroglio of tribal politics led to his increasing and uneasy reliance on an upwardly mobile Amin. By 1971 Amin was major general and commander of Uganda's army. Apparently he also was embezzling army funds and arms to support his own personal army. Within the regular army Amin engaged in a fierce power struggle punctuated by regular assassinations of officers he opposed.

Shortly after leaving for the Commonwealth Prime Ministers' Conference in Singapore the same year, President Obote (a self-proclaimed title following his abolishment of the traditional kingdoms of Uganda) issued a house arrest order for Amin and further ordered him to provide a detailed statement of accountability for his expense of army funds. Obote, a Langi, also suspected Amin of having murdered a fellow Langi, Brigadier Pierino Okoya, one of Amin's military rivals, plus his wife. The house arrest order was a serious misjudgement. So was Obote's concomitant decision made at the urging of Julius Nyerere (president of Tanzania) to attend the Singapore meeting at such a time. After Obote flew to the meeting, his subordinates were so afraid of Amin that, instead of arresting him, they merely showed him the arrest order. Amin immediately mobilized his troops, and with Israeli assistance, smashed the units loyal to Obote. Amid roars of approval from the capitalistic Baganda (Uganda's major tribe) disgusted with Obote's socialism, Amin announced Second Independence Day and proclaimed himself temporary leader of Uganda. With less fanfare, the West sighed in relief at the downfall of another potential Communist. Amin soon set a record for self-proclaimed titles, including Ph.D.'s, and culminating with life-president of Uganda. Meanwhile, Obote returned from Singapore to exile in Tanzania.

The life-president suspended or abolished many civil liberties and issued such frivolous decrees as the outlawing of beards on men or slacks on women. And, as Amin proceeded to eat his job (Ugandan slang for offical corruption) by assuming control of most profit-making ventures in the country, Uganda slid into economic ruin. Citizens at all levels soon lived in constant fear of his predatory minions and his not-so-secret torture chambers. During 1971 Amin "replaced" 6,500 soldiers in an army of 8,500. The following year he expelled nearly 50,000 Asians, many of whom were Ugandan citizens, and confiscated their property and goods for redistribution to his favored soldiers. By 1979 Amin had executed an estimated 300,000 Ugandans (Amnesty International figure), and nearly 250,000 others had fled into exile, many joining Obote in Tanzania to wait for the day they could invade their homeland and turn the tables on Amin.

The current purge (February 1977) was among the most notorious. Our reports filtered in by bush telegraph. The government-controlled Voice of Uganda claimed that a force of Tanzanians assisted by Acholi and Langi rebels under the command of Obote was invading Uganda and attempting to overthrow the government. All Langi and Acholi were allegedly involved; consequently the entire tribal memberships were enemies of Uganda.

These allegations served as a post hoc license for genocide by government troops drawn mainly from West Nile tribes (including Amin's), who were traditional enemies of Langi and Acholi. Members of either tribe serving in any governmental job, including post office clerks or grammar school teachers, were sentenced to death without warrant or trial. Soldiers and inquisitors of the dreaded State Research Bureau (SRB) rounded up suspects and executed them wherever it was convenient. Because Obote was a Langi, this tribe was more hated and feared by Amin, and its members were often beheaded. Acholi were shot. At least one group of prisoners was bound together with wire, doused with kerosene, and immolated. Their charred bodies allegedly were tossed into Lake Nybikere, a small crater lake along the road from Bigodi to Fort Portal.

Another group of prisoners was trucked to a crossing of the Mpanga River, the eastern boundary of Kibale Forest, and were machine-gunned en masse and left to rot. One wounded man crawled from the massacre in the dark to eventually reach Kanyawara where sympathetic Batoro gave him sanctuary. To my surprise, many local people gave sanctuary to the Langi and Acholi lucky enough to have received advance warning of their impending arrest. This humanitarianism seemed to personify Uganda's common people in general. Despite this, an estimated 3,000 people would be butchered during this week.

Despite the army base in Fort Portal, Toro District was far from the mainstream of Uganda's political concerns, so little news reached us. Was the blood-letting as severe as the bush telegraph said? Would they come after Otim?

A new twist developed in the government's account of the invasion: Amin now claimed that American mercenaries were assisting the rebel forces. All Americans were now Uganda's enemies. I consulted my maps for a clandestine hiking route to Zaire, a dubious sanctuary.

Luckily for Tom and Lysa, a few weeks earlier they had flown to Zanzibar to observe a rare subspecies of red colobus. Their houseman, Vincent, was guarding their house. Vincent appeared unexpectedly at Ngogo. He explained to me that agents of the State Research Bureau had arrived at Kanyawara to apprehend Americans, only to find that Tom and Lysa were missing. They then consulted their list and asked Vincent where I was.

"At Ngogo," Vincent replied.

"Where is Ngogo?" one asked.

"It is far into the forest; you would never find it."

"You can take us there."

"I cannot. I am *askari* (guard) here and thieves would rob this place if I left it. Besides, I have never been there," Vincent protested.

Perhaps the prospect of hiking through many miles of jungle and swamp dimmed the ardor of these parasitic and bloody agents of the SRB; I was glad for the first time that Ngogo was located very incorrectly on the topographic map. I was indebted to Vincent for taking the personal risk of parrying the SRB.

Next Amin announced a mandatory gathering at Entebbe of all whites in Uganda. But because the meeting was to take place only ten hours after I heard of it, it was already too late to comply. I would not have gone anyway, preferring my prospects in the forest to those in the real jungle outside it. Our second game guard, Mukasa, returned from Bigodi on the day of the Entebbe meeting and told me Voice of Uganda's latest announcement explained that the meeting had been ordered to present Americans specifically with the option of forfeiting all their personal property (vehicles, radios, et cetera) and leaving Uganda, or merely signing over titles to such and remaining in Uganda. Upon their (our) eventual departure, the property would be retained by the state. No reentry visas were to be issued to Americans.

Meanwhile I had spent most of February searching for new places to observe chimpanzees. The specter of genocide and my possible deportation— or worse—infused my quest with an uncomfortable urgency. How could I find out anything useful about the chimps of Ngogo in the short time I might have left?

And, despite the apes having seemed so numerous a few weeks earlier, they now seemed to have vacated the study area. Pant-hooting drew me beyond trail's end at the southeastern boundary into terra incognita. Because it is so easy to become lost in thick forest hiding a gray sky that itself gives no hint of the sun's position, I left the trails that made orientation so simple and threaded my way along faint, interweaving game paths with a compass and trepidation. I heard something approaching and I stopped.

Exhaling a soft grunt at each step, an adult male chimpanzee knuckle-walked (the first knuckles of his fingers, rather than his palms, bore his weight) toward me. Behind him followed three others. He noticed me when only a few yards away, stared at me in surprise, then bolted north. The chimp directly behind him ran south. The third one, a juvenile male, climbed a sapling and watched me nervously. The fourth spun around and hurried west. I ceased scratching but continued to feel foolish. The juvenile descended and scuttled southeast. It all happened so fast that I recognized none of them.

I continued along attenuating paths, then halted at a majestic *Cordia* tree standing in a glade littered with the trunks of trees felled by a freak wind. I stood directly beneath the blooming *Cordia* and gazed around. The chimpanzees were acting like fugitives. I *was* a fugitive. I scanned the forest surrounding the glade until I was certain it held no apes. I felt like sitting down for a change and glanced around me for a dry place. Shrieks suddenly split the air so close that I flinched. A shower of minute pale green blossoms rained on me. I craned my neck to see four chimpanzees pumping the limbs of the *Cordia* up and down in a "branching" threat display and screaming directly above my head. More blossoms rained upon my face. I could not help feeling embarrassed at having been so inobservant. Unknowingly, I had approached too close to them. They immediately quit snacking on blossoms and abandoned the *Cordia*.

I continued and eventually found another large *mucuso* with ripening fruit. I settled in the best spot for a view that the terrain offered, unfortunately a niche in the understory nearly ninety yards away across a creek. I sat there and tried to forget about Idi Amin Dada and twentieth-century Africa.

A marsh mongoose ambled down the creek and halted a yard from me. It snatched an orange freshwater crab and, oblivious of me, made a meal of the unlucky arthropod. It cracked open each leg and nibbled the meat, then it examined the remains to insure nothing edible had been overlooked. Meal over, it continued down the creek not having noticed me almost within arm's reach.

The first chimpanzees to visit the tree ate little because the figs were too green, but they stayed to groom one another and loll about. Eventually an adult male moved into an adjacent tree, a *Pterygota*, and tore off one limb after another. I strained my eyes to make out why. Was he building a nest? I focused my field glasses again and watched as his strong fingers and teeth peeled off a long strip of bark from a limb. He ate the strip and chewed with exaggerated vigor, as if it were rubber. Within a few minutes an old female with her infant, plus a subadult male and a juvenile male joined him in tearing off limbs. Soon two more adult females with infants and a juvenile climbed into the *mucuso*, then the *Pterygota*, and assisted with the pruning for another hour.

When I first learned to identify *Pterygota*, a tall, straight-boled, big-leafed tree that produces no succulent fruit, I never suspected chimpanzees would eat any part of it. The tree's most notable attribute was that its leaves produced a passable substitute for toilet paper, important because the real thing was not available. Now branches were snapping throughout the crown as a half dozen apes ate the inner bark like popcorn. Over the following seasons I learned that, far from being ignored, *Pterygota* was an important tree to chimpanzees. They built comfortable nests in them, and, in addition to the inner bark, they ate its young leaves, flowers, and the succulent immature wings of its seeds which were bunched together in apple-sized pods. Some chimpanzees spent entire days in one or two *Pterygotas* and wreaked havoc with the hapless plants' attempts to reproduce.

An hour after the last apes began stripping bark, all of them except the subadult male descended to the ground in the gathering dusk. The subadult shifted to a new position to observe me as I started the two-mile walk to camp. I pondered my ignorance concerning the complexities of foraging, and felt acutely the magnitude of the challenge of documenting it accurately.

Eating is universal among animals, and food is revealing. What an animal eats, and when, where, and how it eats, especially with respect to the choice of foods available, have generated thousands of scientific papers. Many analyses of the spectrum of foods available to primates, their nutritional needs, plus their abilities to harvest and digest potential foods, have been aimed at discovering principles of feeding ecology. A major hypothesis hovering at least behind the scenes in most such analyses is that primates attempt to optimize their foraging so as to harvest the greatest benefit in nutrition for the least expense in energy. Like thrifty humans shopping in a supermarket, they should habitually attempt to pay the least to eat the best.

But optimization as a goal in foraging is complicated by several factors. For instance, if a primate is at increased risk of being killed by a predator while foraging, it can adopt several strategies to reduce this risk: keep its foraging time to a minimum by concentrating on high value foods, forage at night when fewer of its predators are present, forage in a large social group conferring greater immunity to being preyed upon (due to the geometry for the selfish herd), or forage in an open, or otherwise protected habitat which may be safer though less rich in preferred foods. Any of these strategies can sabotage an attempt to analyze optimality. And the odds are that none of them would even be recognized as antipredator strategies. But if a scientist gets around this somehow, a gamut of further complications awaits.

Foremost among these complications is the ability of a primate to harvest or process available food. For instance, many hard-shelled fruits and nuts exist in primate habitats but the inability of most species to crack them, like the inability of redtails to gnaw through *Monodora* fruits, forces them to concentrate on other less desirable foods. If a primate cannot harvest a food, or once harvested, cannot process it, it is not genuinely available. Such complications make it difficult for an ecologist to demonstrate cases of absolute optimization.

Still, the *tendency* to optimize nutritional intake for time and energy spent foraging is apparent in most species, despite discrepancies between what is available in the environment versus what is actually harvested. But another major discrepancy arises due to the difference between the nutritional *content* of a food versus it nutritional *value* to the forager. Primates require carbohydrates, essential fatty acids, proteins, essential amino acids, vitamins, and trace elements. Many requirements are known from research on humans; the physiology of primates, especially the great apes, is very similar to ours, though some discrepancies arise when comparisons are made with wild apes. For instance, international health organizations use a standard minimum of 0.016 ounce of protein intake per pound of body weight per day for adult humans. Gorillas studied by Alan Goodall in Zaire, however, averaged an intake three to six times higher in protein per pound of body weight per day. Unlike foods marketed in the U.S.A., the nutritional contents of wild foods are not printed on the package.

But the apes forage almost as if they were. They seem to have a physiological body wisdom that communicates their nutritional needs and, further, understands and remembers the nutritional values of different food types. This body wisdom can help provide indirect evidence concerning the difference between nutritional content and value. For instance, most plants contain carbohydrates as structural cellulose or lignin which the digestive tracts of chimpanzees cannot process—but bacterial symbionts in the stomachs of red colobus, or even gorillas, can. Hence, a plant with fixed nutritional content can have extremely unequal value for different primates.

Sometimes the plants themselves provide signals to foragers to announce the presence of nutrients. This sounds altruistic, but once again it is merely enlightened self-interest. For example, lack of ripeness of fruit normally coincides with the seeds within being unready for a consumer to ingest for subsequent dissemination. Highly edible fruit does not appear on a tree by an accident of evolution: the tree produces it at some metabolic expense to enlist by reward animal assistants who, many hours after eating the fruit, inadvertently drop the plant's seeds some distance away. The fruit's visible change to ripe color accompanies both the now nutritious bait-reward and, what is important to the tree, mature seeds. The ripe color is equivalent to a neon sign saying, "EAT HERE!"

By complete contrast, many other trees wage constant chemical warfare against herbivores (plant-eating animals). American botanist Gerald Rosenthal listed an amazing arsenal: plants not only produce repellant chemicals and lethal toxins, some mimic the growth hormones of herbivores and cause them (insects usually) to develop into nonreproductive freaks; others mimic the herbivores' alarm pheromones, the chemicals the animals produce to warn other members of their species, prompting them to flee rather than eat; still other plants mimic hormones inhibiting herbivores from maturing, so they die without having reproduced. It seems most plants contain a chemist's catalog of alkaloids and other complex molecules produced at some metabolic expense which cause mammals to become ill and vomit or even die. Such poisons usually increase in concentration as the particular plant part in which they are found matures, hence the marked preference of most primates for very young leaves (which also contain more usable proteins). The Sitka willow of North America has perhaps the most surprising ability to thwart predators. It waits until it actually is attacked, then sabotages the nutritional quality of its leaves, and simultaneously *communicates* somehow in a heretofore unknown way to neighboring willows that have not been attacked to do likewise; at the same time their nutritional quality also plummets.

In plant communities with limited nutrient resources, such as tropical forests, slow-growing plants tend to be more successful in the long run than fast-growing plants. Botanists Phyllis Coley, John Bryant, and F. S. Chapin III reported that such slow-growing plants also invest significantly more energy in producing poisonous alkaloids, tannins, and phenols, to make themselves unpalatable. The burden is on the foragers to either evolve immunities to these toxins or ways to neutralize them. Among fast-breeding insects coevolution of metabolism to combat these defenses is common. Bill Freeland and Daniel Janzen reported a variety of intestinal adaptations of the more slow-breeding mammals, but admitted that their primary strategy is simply to forage on mixed species to cut down on the dose of any single toxin. Sometimes, however, these toxins are what the primate wants most.

In a report written after my research at Ngogo, Toshisada Nishida and Richard Wrangham described a puzzling habit of some Gombe chimpanzees seeking the leaves of *Aspilia* immediately after leaving their night nests at dawn. Rather than chew the *Aspilia*, the apes rolled the leaves in their mouths and swallowed them whole, a unique process for any animal eating leaves. Nishida and Wrangham's initial analysis indicated that after swallowing the *Aspilia* leaves, bouts of feeding seemed longer than average, a clue that *Aspilia* contains desirable stimulants. This possibility led to equating the chimpanzees' habit of making a beeline for *Aspilia* in the morning with the human habit of groping for a cup of coffee before attempting to think. But upon reanalyzing the lengths of feeding bouts, Wrangham and Nishida found no significant difference. So they started looking for other clues for the apes' strange craving for *Aspilia*.

Their biggest clue came from local Africans who used the same two species of *Aspilia* in medicines for a wide variety of ailments. Wrangham took leaves to a couple of laboratories in the U.S.A. for analysis of their biopharmaceutical properties. The biochemists excitedly reported that the *Aspilia* contained very high concentrations of thiorubrine-A, an antifungal, antibiotic, antihelminth, and even antiviral compound at least 200 times more powerful than DDT, which, and this is the serendipitous part, also had a low toxicity to mammalian cells. Although the potential of *Aspilia* in human medicine is unclear, the biopharmaceutical analysis of this plant tells us a great deal about its important role in the self-administered medicines of chimpanzees. *Aspilia* also provides an excellent example of the difficulty in determining why chimpanzees seek out and eat certain plants in preference to others.

So optimality is possible, if ever, only after a population of primates has determined what, from literally hundreds of choices, is not only *good* to eat but what is *best* to eat. And the only way for an ecologist to measure optimality is to assign values to certain foods based on some relationship such as energy and perhaps nutritional *content* of a food minus the energy spent in harvesting it. This is obviously oversimplified because a thorough consideration would have to include the nutritional and medicinal *values* of each food, the cost of digestion, the adverse effects of plant toxins, the risk of being preyed upon while harvesting the food, and so on. This may go without saying, but I was a long way from accomplishing anything like this with the chimpanzees at Ngogo. And, for reasons which should be clear, a complete analysis of optimality in diet has never been accomplished for any rain forest primate. Despite the unattainable nature of a full documentation of optimality in diet, most ecologists agree with the general conclusion that, in a world of limited resources being sought by animals in competition with one another, the winners will necessarily have been more efficient in their foraging than the losers. And because foraging ecology is still the most revealing of the underlying limitations of primate social structure, I had to document as much as I could.

Obviously, the total diet of a population, or even community of chimpanzees, evolves over generations and consists of a set of historical decisions maintained by mature apes demonstrating to young ones by example which foods are good to eat. Another comparison made after my research by Toshisada Nishida, Richard Wrangham, Jane Goodall, and Shigeo Uehara of the diet of Gombe chimps compared with that of chimps 100 miles away in the Mahale Mountains indicates the influence of chimp history and the arbitrariness of chimpanzee diets. Athough the diets from both studies were similar, of 143 species of plant foods common to both regions, only 85 species (59.4 percent) were eaten in both areas. When considering the parts of the plant, or food *types*, the diets were even more divergent. Of 286 food types available in both regions, only 104 (36.4 percent) were eaten in common. At first glance such great differences seem strange and unexpected, as if, instead of being optimal, foraging *is* arbitrary.

But Nishida and his colleagues speculated that many of these might not in fact be differences because of the greater effort spent at Mahale to investigate food types, because the Gombe food list may be inflated by counting species outside the territory of the apes observed, because of listing plant parts tested only rarely by the apes, and because of listing other foods eaten only a few times (in desperation?) but normally ignored. Despite these extenuating aspects, major differences remained: for instance, chimps in one region routinely pounded open hard-shelled fruits while apes in the other apparently were ignorant of that technique; conversely, the nonpounding apes regularly "fished" for *Crematogaster* ants in their nests in tree boles while the other chimps ignored them completely. These and many other differences leave one with the suspicion that several techniques to process foods, or even to recognize foods, were "invented" by one population but not by the other, or perhaps the second one lost techniques through forgetting. Significantly, in both regions only immature chimpanzees innovated by exploring possible new foods. Once again optimality takes a back seat, this time to the inquisitiveness and intelligence of the foragers—and to the communication of their discoveries.

Even though perfect optimality may never be attained by foraging chimpanzees, their reproductive success does hinge on good nutrition. This is obviously true for females, who nearly always are eating for two: during pregnancy, then later during three or four years of lactation during which they also carry their infants. But, if the hypothesis is correct that the reproductive success of male chimpanzees is dependent on a complex social network of their relatives for maintaining their territory through combat, then they require optimum nutrition as much as any professional football player. In short, if they were the refined killer apes of the hypothetical model, they *should* attempt to forage optimally. Were the apes of Ngogo doing so?

This important question fixated me on observing foraging. Conveniently, because chimps at Ngogo spent from 52 percent (females) to 62 percent (males) of their daylight hours foraging, they gave me plenty to think about. One problem arose due to my vigils at fig trees: my decision to station myself at what I *knew* to be hot spots of foraging made my observations there nonrandom and, hence, inadmissible in analyses of proportional representation in the chimps' diet. In other words, because I was tracking the food instead of the forager, my perspective allowed me opinions on how important chimps are to figs but not how important figs are to the apes. Intuitively, of course, the repeated, prolonged bouts of fig eating by chimpanzees convinced me that they were important. But beyond intuition, foraging data from my vigils were useless.

Fortunately, my time at these vigils brought familiarity and tolerance between the apes and me. And as my project progressed I was increasingly able to make random observations of chimps traveling, resting, and foraging wherever I found them. Eventually my analysis of 600 random observations of foraging on 51 food types (fruit, seeds, blossoms, bark, cambium, wood, leaves, and animals) both vindicated my intuition and presented an ecological conundrum. During 78 percent of their time spent foraging, the chimpanzees were eating fruit.

Modern ecological models of foraging almost universally agree that large mammals must concentrate their foraging on common, easy-to-find foods—both because large mammals require a lot of food and because their ability to extract adequate nutrition from common foods is superior to that of small mammals. Because small mammals require high quality but need only a little, they can afford the time required for searching and are usually ecological specialists. Chimpanzees are definitely large mammals. But their primary food is one of the rarest commodities in the forest, one competed for hotly (see Chapter 9). And not only were they specializing in a rare food *type*, my research further revealed that even the *species* of trees they sought were significantly among the rarest in the forest. According to current ecological wisdom, chimpanzees should not exist!

Rule breakers though they may be, their life is not easy. The scarcity of fruit and the difficulty of being the first ones to it when it ripens is a challenge on two fronts. First, as I mentioned in Chapter 3, they must possess superior intelligence, time sense, and spatial memory. Even so, physical superiority may be an essential attribute in situations where the apes arrive to find monkeys already exploiting a tree. The second main challenge is more complex and everchanging. In fact, it is the greatest anomaly in the existence of chimpanzees.

Chimpanzees are social. They are so social, so gregarious, and so expert at social communication (for example, long-range communications of identity, greetings, reunions, reassurance, mutual grooming, dominance and subordination, food soliciting, and even food advertising—many of these are discussed later in this book) that they appear to some scientists to be inexplicable examples of social overkill. But nothing could be further from the truth. Their social talents have evolved to combat the constraints of ecology.

The distribution of fruit is ephemeral and patchy, and an average fruit tree normally provides food for only a few adults. For instance, a tree good for an hour of foraging for three adults would provide only six minutes each for the thirty adults of the Ngogo Community. So, if the community were to travel as a cohesive unit, it would need to visit ten times as many trees and travel several times the distance in a day as the party of three. An average party at Ngogo contained three individuals who spent 10 percent (females) to 12 percent (males) of their daylight hours traveling already, so escalation of travel needs would quickly detract from the time when the apes normally eat, rest, and engage in all other activities. The apes' solution to these ecological constraints is community fission into small foraging parties that make it feasible to subsist on scattered, expensive, and rare resources. In fact, in terms of energy spent for foraging, a solitary ape is the most efficient. The anomaly is that they have beaten these serious ecological restraints on social structure not by becoming solitary, as their cousins the orangutans have done, but by evolving diverse and sophisticated abilities to reform social bonds so parties can periodically fuse into a community again.

Beyond mere social talents, males, who are responsible for most long-range calls and more traveling, regularly give food calls at large fruit trees and actively reward with easy food parties who arrive and fuse with them. This may tend to upgrade the nutritional intake of the community as a whole, though admittedly probably not by much. But a little may go a long way. Chimpanzees raised on lab diets in captivity generally mature years sooner and attain adult sizes as much as 50 percent larger than wild apes. This is clear evidence for highly nutritional food being limited, at least seasonally in the wild. Consequently, each increment toward superior nutrition can be important to a chimp mother, and, conversely, poor nutrition can prove disastrous to her reproductive career.

A corollary of this is that females should not make unnecessary metabolic demands on themselves. In this light it makes sense that they travel less than males. A female actually needs to travel only far enough daily to insure that she and her dependent offspring receive good nutrition. Oddly, males at Ngogo spent 10 percent more of their daylight hours foraging then did females, who on average rested a full hour more of each day. These differences in activity budgets suggest an incompatibility between the sexes. These differences also pose the key question: what is the payoff for males going to such trouble to reform social bonds, to regularly reglue the community together? Their elaborate social adaptations to combat the constraints of ecology demanded a good explanation.

At this point I suspected the answer was that their adaptations for social cohesiveness were strictly to insure the solidarity these males required when in combat with alien males over territorial disputes. But also at this point I was far short of having enough data to assure myself I was correct. Now I was certain only that Ngogo chimpanzees were specialist foragers in contradistinction to what is known about optimality. Metaphorically they had thumbed their noses at ecology and, against the odds, had evolved a fusion-fission social system allowing each of them to forage somewhat optimally on rare resources distributed so patchily as to be prohibitive for large, group-living mammals. Yet they were one of the most richly social primates known.

Currently the best data available to me on territorial killing came from Idi Amin Dada's genocidal purge against the innocent Langi and Acholi. His rhetoric directed against American mercenaries directly threatened me as well. Amin was still incensed about the embarrassing success of the Israeli commandos who had rescued their compatriots from a hijacked jet grounded on his Entebbe airstrip less than a year earlier; he even ordered a hospitalized woman from the flight to be gratuitously murdered for spite in frustration at the Entebbe raid. To avoid being anyplace where his State Research Bureau might show up, I continued to spend my days in the vicinity of the large *mucuso* intertwined with the *Pterygota*.

The figs finally ripened and were harvested by chimpanzees. At one point twenty-four of them foraged in the tree simultaneously. But I found that being ninety yards away was a handicap because I recognized only one chimp I knew, an adult male I named Notches for a single bite-in-the-apple notch in each of his ears. I finally had to admit to myself that I was too far away, but no closer vantage point afforded a view of the tree.

While mulling over my file of individual identities I heard grunts, snuffling, and heavy breathing closing in on me from behind. As I quietly stood to face the mysterious hazard in my imagination I pictured myself being squashed by an elephant or buffalo. Less than three feet above the ground, however, a pair of bat-wing ears waved as they advanced toward me. A shiny pig snout materialized between them. The boar veered left to pass me only a body length away. I had stood so slowly that he never recognized me as an animate creature. All oblivious of me, eight more bushpigs sows, immatures, and piglets, followed him with their snaky tails wriggling.

This was a bonus: bushpigs were not commonly active during the day. Struhsaker sighted them only about twenty times during nine years in Kibale, and I saw them only ten times. My encounters were always at close quarters, and usually the animals were unaware of me. It was easy to see how the piglets could become the prey of chimpanzees. Bushpigs sometimes grunted deeply as they traveled, then stopped to roll on the forest floor and rub their warty faces on the earth. They cropped new vegetation along the path as they walked. Sometimes they shredded rotting logs and tangled vegetation with their tusks to feast on the grubs and other fare in the litter. They rumbled down creek courses like miniature bulldozers, shoved boulders aside with their snouts, and gobbled the unfortunate creatures once happily esconsed beneath. But they rarely looked around and noticed me, even when twice I carefully stalked close enough to slap them on the rumps. Bushpigs continued to alternately scare me unseen with their sudden sonorous grunts and to entertain me with their oblivious attitude. It continually mystified me how they could be so lax about predators and yet survive.

I climbed the hill to camp and found Simon, just returned from Fort Portal on a resupply trip, and an attempt to determine how Americans stood with Amin. He looked at me, paused, maybe for dramatic effect, and said, "There were still bodies along the road . . . they were getting pretty high."

"Did you make it to the Governor's Office?" I asked.

"Yeah. Talked with a bloody clerk. The official word now is it's business as usual in Uganda. Amin's meeting in Entebbe with us whites never came off." Simon paused for a moment, "They killed Ogwal."

"They murdered Charles Ogwal, the Provincial Secretary? The bastards. What about the invasion?"

"The whole bloody story was just Amin's excuse to wipe out some more Langi . . . and to rip off you Americans."

"It seems like people in Uganda accept all this pretty fatalistically," I said. "It's almost as if their history of tyrannical kings has injected fatalism into their blood."

"They're just uneducated peasants. Democracy is a new concept to them," Simon continued. "It will take them a long time."

"Especially with Amin in charge. Did you have trouble getting petrol?"

"No. Kenya lifted the embargo."

In March I found another *mucuso* perfect for observation. I was able to sit in the open only forty yards from the crown. I frightened several chimpanzees away at first, but by holding my post I not only added considerable new detail to my identification file and gained acceptance by several more apes, I began developing relationships with individuals which reflected their personalities. Some of these lasted for years.

My reunion with R.P. occurred here. He usually traveled with males, some of whom I did not recognize: a subadult I named Newman, a prime male I named *Nane* (Swahili for number eight since he was my eighth male), and a magnificent, massively muscled, long-haired male, Eskimo, who dominated all the others. Eskimo initiated wild and wailing pant-hooting sessions during which he vigorously swaggered through the tree, then sometimes brachiated (swung by his arms) from limb to limb like a gibbon hooting all the while. The other males chimed in like choir boys.

R.P. started using me here at this *mucuso*. When first climbing up with other males he consistently rushed along a limb toward me and slapped its smooth bark loudly with his open palms in a display of bravado directed at me. In different ways his performance was good for both of us. I had the distinct impression after R.P. threatened me and I simply scratched myself and showed no aggressive response, that the other males became far more relaxed in my presence. By contrast, when R.P. arrived at the *mucuso* and no other males were around, he simply glanced at me and proceeded to forage seemingly without a care in the world. It did not take me long to understand that it was irrelevant to him whether I was impressed with his displays, or even noticed them, they were intended to impress his male companions. And they did. Months later, after he had worn out the display-at-me gambit, he demonstrated the lengths to which he was willing to go, simply for appearance, to maintain solidarity with other males.

He and four other males were foraging intently in a *Ficus exasperata* during a season of very little fruit. So intensely were they stuffing themselves with fruit and scanning nearby boughs, as if trying to decide which new limb offered them the absolute maximum quantity of figs before a neighbor beat them to it, that none of them had noticed me standing less than forty yards away. One male, a stranger, finally spotted me after half an hour and hastily descended. A second male, also a stranger, glanced around for the source of the first one's fright, saw me, and followed him. The remaining three, including R.P., saw me, then literally hurled themselves out of the *exasperata* as if chased by the devil himself.

I knew R.P. had seen and recognized me, and it bothered me a little that he fled like that. But two hours later he returned alone, climbed part way up the *exasperata*, then, as if he had memorized my exact location, calmly glanced at me, then scanned the crown and climbed up higher to forage for forty minutes as if I did not exist. For a full five minutes after he quit plucking figs he sat as close to me as the tree allowed and casually observed me, as if hoping to add to his store of information. Then, equally casually, he descended into the dense understory and was gone.

I am convinced that R.P. had known I had posed no threat earlier when he had fled with his companions, but had decided that behavioral solidarity was the highest priority of the moment. I suspect that upon reevaluation R.P. decided that abandoning the *exasperata* and its rare load of figs need not be permanent. Hence his return. This type of sacrifice for a social priority was a characteristically male behavior at Ngogo; it would have been completely out of character for a female habituated to me to flee with other females who feared me. Later, as our relationship grew firmer and he aged and gained in status, R.P. ceased using me and actually became attracted to me.

Gray was another chimpanzee with whom I would have many encounters during my study. She was an ancient female, apparently postmenopausal, missing a finger, and carrying scars from an apparently eventful life. Her calm but inexplicable acceptance of me contrasted with the exaggerated nervous reactions of the pretty young female Zira, who was her companion nearly half the times I saw her. Zira could barely steel herself to remain in my sight. Whenever I made a minor shift, say from sitting to squatting, Zira rushed headlong for an exit tree and flew down it. But because several other chimps were at the same time ignoring me completely, Zira soon climbed back up and resumed feeding, while keeping one eye on me the entire time. My slightest move would send her out of the tree again. It was tempting to twitch simply for the effect, but I resisted. Eventually Zira became as blasé as Gray, who acted as my primary informant concerning the priorities of female senior citizens at Ngogo.

Often, when male chimpanzees were not in the *mucuso*, mangabeys, redtails, blue monkeys, baboons, and/or red colobus were. Frequently different species fed here together. The apes occasionally chased monkeys out of the tree, probably to reduce their consumption of figs, though sometimes seemingly just for sport. Soon I assumed that all chimpanzees were dominant over monkeys, but one of my other primary informants on chimpanzee life showed me I was wrong.

Owl, a young adult female chimpanzee, climbed into the *mucuso* alone. The tree already had two dozen mangabeys and redtails foraging in it, none of whom reacted to her. As Owl climbed, she paused to study the approach of some red colobus about thirty-five yards southwest. She plucked only a few figs, made a quick circuit to the northeast section of the crown, then sat directly above an exit tree.

Meanwhile an adult male red colobus had entered the *mucuso* and had followed her halfway through it. He vocally threatened her with "chists" (a red colobus threat vocalization) but paid me no attention. His threats quickly escalated to advancing toward Owl in short, abrupt pairs of steps. After several aggressive bunny hops the monkey rushed toward her and she dropped six and a half feet lower in the crown. The colobus continued feinting lunges and threatening her for several seconds, then rushed her with teeth bared in a snarl. Owl dropped another ten feet, but the male launched himself through the air after her. Limbs snapped as Owl screamed and swung quickly to the ground. The monkey followed like a kamikaze pilot bent on exploding the enemy on contact. Owl beat him to the ground and loped away at high speed. The male climbed back up and gradually calmed down. This episode seemed bizarre to me, but Owl's behavior suggested that she more or less expected it. Soon I witnessed so many competitive clashes between monkeys and apes (see Chapter 9) that I grew to expect it too.

Owl was so accepting of me that I decided to use her to experiment with my methods. Originally I had hoped as the apes became tolerant of me to follow them from one feeding patch to another. But so far each time I had tried this, they bolted through the understory and left me standing in a seemingly empty forest. The chimpanzees could travel much faster along their tunnel-like game paths because my extra three feet of height refused to slide through the tangled vegetation.

As Owl dropped again from (another) fig tree and knuckle-walked into dense understory, I followed. A couple of minutes later, after congratulating myself on my stealth, I was embarrassed to look ahead and see her sitting in a low tree a few yards away with her chin propped on her crossed wrists and studying me. I soon discovered that she felt completely secure whenever she was in a tree, even a low sapling, but being on equal footing made her nervous. Many more months would pass before she let me break the ground barrier.

She seemed to concentrate her travels in the core of the study area and the region south of it, so I eventually spent more time in her company than in that of any other chimpanzee. Owl became virtually nonchalant and taught me a lot about the life-style of young adult females. For instance, unlike R.P.'s first priority of social solidarity, Owl's primary concern was food. Socializing was second priority, though she became obviously excited over the prospects of traveling and grooming with other females when food was plentiful. Her extroverted, unconcealed joy over being with preferred companions was excellent testimony to the blessings versus the conflicts inherent to the apes' fusion-fission social system.

The smoke was now clearing from Amin's latest butchery, so my prospects of remaining in Uganda looked good. And by now I had developed working relationships with several chimpanzees of both sexes and mixed ages. Now I figured that merely continuing to find them would be my last major challenge to documenting chimpanzee ecology. But I was wrong. I foolishly had underestimated the influence of the most voracious predator on the planet.

—— 5 ——

The Creature that Ate the Forest

The morning sun climbed above the forest as I followed trail G where it cut close to the southwest island of elephant grass. A hot wall of sunlight beyond the interlacing of drier edge trees contrasted with the deep greens of the forest downhill. The world beyond the edge of the forest was different, different enough sometimes to seem like another planet.

I crouched low to thread my way along the last bit of game trail connecting G with the grassland. Abruptly my disintegrating hiking boots had granite beneath them. I knelt to touch it. In the forest, where mud and rotting vegetation were the norm, clean dry rock was a novelty. Tall grass surrounded the slab as if impatient to obliterate the gap. In its center was a bull's eye of baboon dung. I climbed higher and savored the solidity beneath my feet. A camouflaged bird, a nightjar, fluttered from my tread at the last second as I crossed onto another level slab of granite crowning the low hill.

In the west sunlight glittered off the glaciers capping the Mountains of the Moon. For months these equatorial rivers of ice had been hidden more than three miles above sea level by a perennial shroud of clouds. But now the sky was blue and I was seeing them clearly for the first time. The Ruwenzori were magnificent; I felt a compulsion to start hiking toward them.

It was easy to resist. Two days earlier, in the fading light of dusk, I had been hurrying through the understory off the cut trails when I slipped on a mossy rock and lacerated my knee. Now, to minimize the distance I would have to limp, I found a thronelike block of granite to sit on. I would wait for chimpanzees to pant-hoot, then take a compass fix on them. This throne also offered a panoramic view of the southern extension of the forest.

Seclusion in the forest had faded my skin to ghostly pale. I stripped off my clothes to synthesize vitamin D. For two hours I listened for pant-hoots and turned pink. When I pulled my fatigue pants back on I noticed a painful bulge on my big toe, my first chigger egg sack. Chiggers are minute spiderish creatures that scurry along the ground but use mammals to breed. The female injects her ovipositor into the skin, especially around toenails, to deposit dozens of eggs which grow to maturity, and, if not removed, can hatch into a disgusting mess. My rotting boots were allowing chiggers in. I excavated the squirting sack with my Swiss Army knife.

No pant-hoots. Again I scanned with field glasses but saw only redtails and red colobus foraging along the edge of the forest. I pulled a sock on. The alarm call of a bird rang out about twenty yards away on the far side of a clump of small trees. Through their branches I saw human forms approaching me through the tall grass. At first I suspected that Otim and Mukasa were patrolling here, but when I counted four of them I knew they were not on our team.

Dressed in ragged, earth-colored clothing and carrying a spear, *panga* (machete), and basket hanging from a shoulder strap, the first poacher climbed onto the slab to walk past without seeing me about fifteen yards away. I yanked a boot on. The second man, also carrying a spear, had barely passed when his glance caught me, pink and attempting to pull on another sock and simultaneously muster an air of authority. He mumbled something to the one ahead, then turned and rushed back the way he had come, almost knocking over the third man following him. The first one returned, looked around, and spotted me. His eyes grew round, then he sprinted after his companions. The tall grass waved and danced as they leapt back down the path. I watched them go and wished in vain for the leverage of a Colt Government .45; by firing it in the air I might have arrested some or all of them. My knee would not allow me to chase them, and, in Amin's Africa, I could not pack a .45.

I returned to camp to alert Mukasa and Otim but found only Yongili. "Where are Otim and Mukasa?"

"They are on patrol, on the Mwengi Side," Yongili acted a little surprised to see me back in camp so early in the day. He considered me eccentric because I usually was out all day. "Why are you looking for them?"

"I just saw poachers, about a half hour ago, down on the southwest island," I answered, "You can see from here . . . just there." I pointed.

"Poachers? Here?" Yongili became visibly nervous. I had never seen him look so concerned. He was young, really, I reminded myself, even his face seemed to hold the innocence of childhood. And Simon had told me that when Yongili had first arrived at Kanyawara looking for work, he had been too terrified to enter the forest. Yongili was small, maybe five feet seven inches and built sparely and, like most African men, seemed to be completely lacking in subcutaneous fat. (Whether this was diet or climate I was not sure, but by now I was also lean.) But these days Yongili regularly cut trails in the forest, even alone, which revealed courage most local people lacked. Most trail cutters, and even many poachers with weapons, avoided traveling in the forest alone at all costs. So why was he so nervous now?

"Yes, there," I told him. "But now they have run south."

"This is bad," Yongili concluded dismally. "Last night I dreamed that poachers chased me all the way to Kanyawara as I was going on my thirty-day leave tomorrow. And now they are here. . . ."

"Not a good omen, is it?"

"You know," he explained soberly, "the last poacher we caught was beaten badly."

"Who beat him? Mukasa?" Mukasa was the only one in camp who, I imagined, possessed the right temperament to seriously chastise a captive. Years ago, during a major intertribal conflict between the Banyoro and Baganda, he had been captured and tortured. He was experienced.

"No! I did," he answered proudly. "He was poaching in the study area."

Despite this omen, however, Yongili went on leave to Bamba and his dream did not come true.

Soon I returned to trail G near the island to continue my dual project of mapping fig trees and counting and mapping chimpanzee nests. Knowing the locations of fig trees helped me find apes (which I had been unable to do lately) and to estimate the densities of their food patches, but the nest count was just as important.

Although some chimpanzees occasionally reupholstered an existing nest for a second night's use, generally each ape beyond the age of weaning (at about four years) built a new one every night. About five minutes of bending in and interlocking several supple limbs in a tree crotch produced the framework for the simian version of a penthouse. After a brief trial period, the designer often sat up to reconsider and then pad the unsatisfactory portions of the basic structure. Sometimes the quest for the perfect bedroom resulted in twenty minutes of leafy additions. Occasionally chimps also spent a minute or less to construct a no-frills nest for a brief nap during the day, but these sometimes became unrecognizable as nests the moment they were vacated, and normally they were distinguishable from night nests. Why is this important?

Taking a physical census of chimpanzees in a large forest is fraught with many variables and requires dozens of repetitions of counts along the census route before it becomes somewhat reliable. And for a solid understanding of their ecology I needed the most reliable estimate of their population density I could make. Their nests are easy to count if one searches slowly and carefully and so stable that only one count per route is needed. I observed the natural disintegration of several nests and discovered their lifespan to be 111 days. And, eventually, I covered 80 miles of rubber-necking at a snail's pace to count hundreds of nests which I used to estimate the population density in Kibale Forest: 6.8 chimpanzees per square mile.

By contrast, the estimate produced by my census data, which relied on actual sightings of the apes, was 3.9 chimps per square mile. The discrepancy results most likely from my not seeing every chimpanzee in my census strip, and, on the other hand, from counting a few day nests as night nests. An average of both methods, 5.2 apes per square mile, is perhaps the most useful estimate. Any way one looks at it the apes were needles in a haystack. Their numbers were similar to those in Gombe, though greater that at the Mahale Mountains. But these figures would be calculated in the future; just now I had been walking, searching, mapping, and recording each nest and fig tree for so long that my mind was wandering.

From uphill, less than 100 yards away where the forest and grassland met, branches snapped and feet pounded. Something was coming in my direction.

Yesterday Mukasa and Otim had found a wounded African buffalo during their patrol. During their attempt to determine how severe its injury was they chased the animal back and forth across the southwest island. Because a wounded buffalo can become dangerous to innocent people, they half expected to have to shoot it. But neither of them was experienced at hunting big game and no shot was fired. The evening sun set on two tired game guards and one aggravated buffalo. Last night they talked of trying again this morning.

Of course I imagined the crashing was the buffalo, but I reminded myself that the periphery of the island was at least a mile and a half long. It would be too great a coincidence. Most likely this was a bushbuck or even a baboon. But, just in case, I looked for a tree to climb. Inconveniently, the lowest branches of the trees large enough to offer safety were several yards above my head. I tried convincing myself again that it was only a bushbuck.

Whatever it was, it was heading directly at me. I stepped behind a tree. A buffalo thudded into view, then halted about ten yards short of me. Except for part of my head, I was concealed behind the tree. She ran along a tangent to my tree and stopped again only ten yards away. I had moved simultaneously to remain hidden by the tree, so now neither of us could see the other, but she knew I was there. I waited for her next move. . . . The silence dragged and I had time to remember several stories about how of all the "big five" game animals to meet wounded in the bush, African buffalo are the worst.

Unable to bear my curiosity I finally peeked around the tree.

She was facing me with her head held low. Her horns looked sharp. Suddenly she turned and, limping on her left foreleg, rumbled past me. I glimpsed a bleeding wound in her shoulder that looked like a bullet hole. Poachers. Here they faced only token opposition and had nearly exterminated the big game. The cow continued downslope limping painfully. As I watched her disappear I noticed that my knees felt weak.

Three mornings later Mukasa appeared gasping for air at the hut. He paused to catch his breath, then his normally somber but elfin face creased into a smile, "*Tembo iko pale* (elephant are here)!" Simon, Mukasa and I hiked a mile to the big hill behind camp to meet Otim, who silently pointed at a scattering of gray blotches in the tall grass 450 yards downslope. A distant crunching drifted up to us on the breeze; we could actually hear them snapping and pulverizing the bamboo-like stalks of elephant grass between their four working molars. I focused my field glasses on a moving blur. Two mammoth ears appeared in the tall grass, then a gleam of ivory. One by one I studied the other eleven of them. A dozen elephants, but only one with tusks.

Gazing down the hill at these Pleistocene mammals changed my perspective of Kibale Forest. It was almost as if I realized fully for the first time that this was Africa. As we watched the pachyderms, they slowly wandered to the cover of the forest and, too soon, vanished. Perhaps they knew they were vulnerable to hunters in the open.

While scanning the now deserted edge of the forest we heard a helicopter hidden by haze. This was the only chopper I have ever heard from Ngogo—and it was miles away—but it sounded alien and eerie. Otim, the Langi, dived for the scanty cover of a small acacia. He was afraid State Research Bureau henchmen were coming to kill him, even though Amin's genocidal purge was supposedly over.

His bush did not conceal him at all. Mukasa studied Otim's tiny acacia and giggled, then he looked away as if by removing the source of amusement from his eyes he would be able to regain his self-control and stop laughing at his partner. Meanwhile, Otim grimaced as if the whirring blades in the sky were ringing his death knell. Finally, because the futility of Otim hiding while we three stood in the open seemed ridiculous, Mukasa began laughing quietly and could not stop himself. Tears were creeping down his seamed face when the sound of the helicopter faded away. Otim finally stood up, looking sheepish.

Both game guards expected the elephants to emerge from the forest at dusk because, as Otim put it, "They do not fear poachers at this place." Rather than battle my way through a shortcut in the elephant grass, I backtracked through camp to enter the forest and continue mapping fig trees, counting nests, and searching for chimpanzees. By late afternoon I had found no apes and so returned to camp, picked up the camera loaned to me by the U.C., Davis anthropology department, and climbed toward the big hill. I wondered if Otim and Mukasa were correct about the elephants returning to the grassland. Mukasa grabbed his rifle and fell in with me.

We scanned the area where they had entered the forest that morning. No sign of them now. Suddenly Mukasa said, "*Wawindaji* (hunters)!" I looked around, saw no one, and wondered if I had heard him correctly. Then, excitedly, he pointed to a high ridge more than half a mile away. I stared at it and finally saw eleven dots on the hillside. My field glasses revealed them as hunters sitting in the grass and watching the edge of the forest below. They had not seen us. I counted three dogs and at least one rifle, and, of course, they were bristling with spears. Otim's optimism that his and Mukasa's presence created a circle of immunity around Ngogo was again unrealistic. Poachers in the wake of elephants, even here, was no coincidence.

We walked toward them. Mukasa advised me, "*Piga picha* (take a picture)." To humor him, and to stall in the hopes that the hunters would notice us and flee, I focused on the tiny dots in the distance and pressed the shutter release button. It jammed permanently. We resumed our advance armed with Mukasa's .375 rifle, three rounds of ammunition (all the Game Department would issue), my pale skin, my hunting knife, plus a broken camera.

After a minute or so we stopped again. One poacher stood, shouldered his spear, and slowly walked away to the east to disappear beyond the ridge. We walked closer. This was beginning to feel like some sort of "High Noon" standoff, but I was acutely aware that my holster was filled with bluff.

A second man rose and strolled over the hill to the north, his rifle shouldered and a dun-colored dog at his heels. Their nonchalance suggested that we were an inconvenience rather than an entity to be reckoned serious. At last the remaining nine men arose in unison and crossed the ridge.

Toro District is one of the garden spots of Africa, with adequate forage for cows and goats. Here there was no real need to kill the few remaining elephants, buffalo, and other wildlife. These men were not poaching as a last resort to stave off starvation but to make fast money. Back in their village, each probably had wives working in their gardens. The escalating value of ivory makes poaching even immature animals carrying a few pounds of ivory tremendously tempting, and the poaching has become so wholesale that now many Batoro were eating elephant meat, which used to be considered almost taboo. This unrelenting pressure by hunters was robbing the complex Kibale ecosystem of nearly all its elephants, the forest's key browsers and most important dispersers of seeds. An earlier investigation of the Kibale herd by Larry Wing and Irven Buss, American wildlife biologists trying to unravel the migratory patterns of wild elephants, found they are paramount in regenerating the forest because they wander so far between eating fallen fruit and excreting the scarified seeds. But poaching has nearly severed this vital link in the ecological chain.

Elephant society is matriarchal: the females of a family form an enduring unit led by the oldest, most experienced cow. As males in the family mature, they gradually spend more time on their own or with other bulls. Because they carry heavier ivory, their chances of living long in this poachers' paradise were slim. I never saw an adult bull in Kibale.

The elephants made life easier for me. They are so heavy that they restrict their peregrinations to traditional paths along the natural contours of the terrain, the paths of least gravitational stress. The easy-to-walk, non-Euclidean maze they created at Ngogo was superimposed upon the contrasting north-south and east-west trail-grid system, which was cut without regard for the swamps, hills, creeks, and boulders that make it an obstacle course. When trod recently, elephant paths provided me a pleasant meandering tour of the forest. They also covered parts of Ngogo the trail-grid system missed. Any stroll along one of their paths ran the risk of encountering the construction crew, but I learned to gauge this possibility by the dryness and the direction of the bends of the vegetation they broke and crushed and by the degree of decomposition of the dollops of dung punctuating the trail. A pile of steaming dung always made me more cautious.

The elephants' wanderings emphasized that Kibale Forest Reserve is only a segment of a much vaster ecosystem. To the south and west, the region known as Queen Elizabeth National Park, extending south from the southern tip of the Mountains of the Moon, is one end of this ecosystem. The Semliki Plains, abutting the opposite end of the Ruwenzori in the north, is the other end. Kibale Forest almost connects these two; Ngogo is only a way station along the migration route. During the December to February dry season (now over) elephants from Queen Elizabeth Park regularly migrated up to 100 miles through the Kibale Forest Corridor Game Reserve (50 square miles) connecting Queen Elizabeth Park with Kibale Forest to, and often through, the central block of Kibale Forest and westward to the lower slopes of the Ruwenzori where the greenest herbage was available. Some elephants may have continued to Semliki, for their full migration routes were never untangled. But Kibale Forest was critical in their home range.

That was before Idi Amin Dada launched a ruthless blitzkrieg against them. The life-president banned tourism (Uganda's second-ranking source of foreign exchange before he seized power) in 1973 and sent covert military units into these last remaining sanctuaries. According to Keith Eltringham and Bob Malpas, biologists specializing in East African elephants, during the following two years soldiers slaughtered approximately 11,000 elephants out of a population of 13,000 for their ivory alone, which was worth a fortune. Since then their population has continued to plummet, with the few remaining tuskers suffering the worst.

With the demise of the elephants, new *shambas* spread across their traditional migration routes, some of them illegally. This encroachment and the concomitant increase of people in their old stomping grounds intensified poaching of the harried survivors. (Later, this self-feeding spiral of destruction would receive a near coup de grace during the 1979 liberation of Uganda by Tanzanian forces. The population of Queen Elizabeth National Park would be blasted to near oblivion by Tanzanian soldiers who sold the ivory and meat via Ugandan middlemen. The skittish survivors in Kibale would be thinned to a few dozen individuals, mostly immature and often wounded. The availability of automatic rifles jettisoned by Amin's fleeing troops in trade for food and civilian clothes would hasten their downfall. There seemed to be no group without a rifle-armed poacher on its trail by 1981 and neither sufficient game guards nor equipment existed to enforce an interdiction. But by 1986 this grim picture had turned around due to shaking down, reequipping, and retraining game guards.)

I had been in Kibale Forest now only four months and already events were expanding my focus on the ecology and social behavior of chimpanzees to include the pragmatic issue of their survival—and that of the forest itself. And each passing day increased my anxiety for the future. Batoro poaching elephants and buffalo imposed little direct threat to chimpanzees, but these hunters were only the minute tip of a mountain-sized iceberg. I was alarmed to learn that, instead of being protected, both the apes and the forest were under attack.

Uncontrolled growth of Uganda's human population was the primary problem. Uganda has the fourth greatest population density in tropical Africa, but here it was not simply growth among the Batoro. The industrious Bakiga people inhabiting the Kigezi District, Uganda's southwest border with Rwanda (Africa's most densely populated nation) and with Zaire in the west, had by now occupied all arable land traditionally theirs. After cutting down 95 percent of southwestern Uganda's forests and converting them to crop land (leaving virtually only one forest, the 129 square miles of the Impenetrable Forest Reserve which protects half the world's mountain gorillas and a unique population of chimpanzees), the overflow of Bakiga had to push northward to survive. Because Toro has much in common climatically with Kigezi, many Bakiga found Toro to be a land of milk and honey.

The climate and soil of Toro are superior to much of Uganda, itself called the "Pearl of Africa," and, over many generations, this may account for the easy-going "mañana" credo of the Batoro, for whom intense and lengthy work is an anathema. Crops are so reliable and life so easy in Toro that even the basic family unit, which in most agricultural societies in Africa is an economic necessity, is optional. It is not uncommon for Batoro women to marry and divorce each year or two and have several husbands sire their children, most of whom become absorbed into the women's greater families.

For immigrating Bakiga the Batoro are often pushovers, and though conflicts exist, the number of Bakiga, and even Bakonjo, immigrating from their crowded lands abutting the Ruwenzori continues to increase in Toro. This increasing pressure has created an accelerating demand for suitable new land, of which the supply is limited. Consequently Kibale Forest was becoming ever more valuable real estate. Just how valuable was a question plaguing Struhsaker and others of us now concerned with Kibale's future. Although in March 1977 we only suspected encroachment into Kibale was severe, eventually Karl Van Orsdol, a graduate student just beginning a project on lion predation in Queen Elizabeth Park (see Chapter 13), would return to Uganda in 1982 to survey the damage in detail. The upshot was that 1977 was the peak period halfway through a full decade of illegal encroachment beginning with the reign of Idi Amin Dada in 1971. Initially the local administration allowed Bakiga to settle along the western margin *outside* the forest, but their overflow soon swamped the available land. By the end of that decade Van Orsdol counted at least 7,000 people living *within* Kibale on the grasslands plus 1,800 people within the forest itself, mostly in the southern block of the Reserve. They were so entrenched that the local government had even built schools and other offices to serve them. Their illegal encroachment had nearly severed the southern from the central block of Kibale Forest—a corridor less than a mile wide was all that remained.

Even though I was unaware of the level of this threat in 1977, the evidence of its intensity was enough to preoccupy me with the menace creeping over the edge of the forest. Two major questions haunted me: was this forest island still large enough to support a healthy population of chimpanzees? And, if so, how long could Kibale Forest be expected to survive?

According to my censuses, nest counts, and observations, each chimpanzee community the size of that at Ngogo (comprising at least fifty-five apes) needed approximately ten square miles of forest. The Kibale Nature Reserve contains roughly twenty-three square miles, roughly 70 percent of which is forest. So the nature reserve probably is just large enough for two communities. But even though it is only slightly smaller than Gombe National Park (thirty square miles) it is far too small to maintain a genetically viable population of chimpanzees, which, based on studies of humans, must contain something on the order of 1,000 individuals. But what about the whole of Kibale Forest?

A concensus of sorts exists among tropical ecologists that the minimum size for a self-sustaining forest reserve is approximately 190 square miles. So, even though the nature reserve is too small, the entire unexploited central and southern blocks remaining in Kibale Forest (containing roughly 144 square miles), plus adjoining secondary forest, probably are enough to remain self-sustaining and support several communities containing the hundreds of chimpanzees necessary to maintain a viable population—but only if it is conserved. Unfortunately, the current status of the forest outside the nature reserve is standing timber awaiting government extraction on a 70-year cycle. Even the extra protection for the nature reserve is only a matter of Forestry Department policy, not of law. None of the region is officially identified as a vanishing ecosystem or as critical habitat. So, even a quick glance at the situation reveals that the primary threat faced by chimpanzees is not poachers (who also tend to be encroachers) so much as habitat destruction.

Some officials of Uganda's Forestry Department are genuinely concerned with conserving their limited forests, which in 1964 covered only 4 percent of the nation but by 1986 had been whittled down to 2 percent. The commercial value of its timber, standing gold for Third World countries, has led to other officials sanctioning rapid exploitation for short-term income. Because he is a guest in Uganda, Tom Struhsaker's recommendations as a conservation representative of the New York Zoological Society were often made as delicately as walking on eggs. Despite awkward protocols, he was instrumental in the expansion of Kibale Nature Reserve from an original plot less than a tenth as big and in the designation of 4.6 square miles around Kanyawara as a research plot.

The vulnerability of Kibale Forest as standing wealth and potential agricultural land is directly linked to Uganda's population explosion. Between 1960 and 1980, despite all wars, hegiras, and famines, Uganda's population doubled from about 7 million to 14 million, and it continues to grow. Hence the demand for more farm land. But even if the entire 215 square miles of Kibale Forest were converted to *shambas*, it would satisfy only three months' increment to Uganda's population. Yet, in the eyes of the people around it, the forest is being wasted by not allowing them to cut and farm it.

Due to poorly marked boundaries and a history of extremely weak law enforcement, massive encroachment was now occurring south of Ngogo. And during legal exploitation in the northern block, after the desirable trees have been felled and charcoal burners have claimed the rest, some sections have been replanted with exotic softwoods promising an earlier second harvest. But this softwood monoculture bears as much resemblance to the original forest as a goldfish bowl does to a coral reef ecosystem, and it is a useless wasteland to the original inhabitants of the forest. Because many tropical forest trees live for hundreds of years, and most species are distributed as rare individuals over a wide area, regeneration is a serious problem. In terms of human lifetimes, the regeneration of large blocks of felled forest to their final, "primary" state, even where dispersal agents are able to provide the necessary seeds, is incredibly long. Norman Myers, a British conservationist from Oxford campaigning for tropical forests, in 1984 reported a team of investigators who estimated complete regeneration requires 1,000 years. But in most cases, following clearcutting or poisoning with arboricides, the complex ecosystem simply ceases to exist.

Relatively, however, Uganda has a fairly good record of forest conservation. Most other tropical nations, with Indonesia and Malaysia leading the pack, are cutting their forests at unbelievable rates, usually employing foreign contractors, and selling the wood for foreign exchange to Japan, the U.S.A., Europe, or elsewhere to finance their rapid transition to the material standards of the developed world. The more I learned about this, the less able I was to maintain a provincial priority centered strictly on the chimpanzees in Kibale Forest. The tropical forest as an ecosystem is in extreme danger and its death transcends human economics.

Ironically, in many regions where tropical forests occur no other man-made agricultural system can persist for more than a few years. Often the replacements are based upon temperate agriculture unsuited to the tropics. Because the chemical nutrients of the ecosystem are poor and what do exist are tied up primarily in the plants themselves rather than in the soil (as they are in temperate forests), removal of the plants severely depletes the nutrients. After removal, the soil is extremely vulnerable to leaching and sometimes becomes lateritic (rock-like) after exposure to sunlight. The usefulness of the region then plummets. Amazonia and Central America are exceptions in that laterization is slower, and, because of this, vast forests have been, and are being, cut and burned to create savanna for cattle raising to supply the growing huge demand for cheaper beef in the U.S.A. created by competition between fast-food chains and the producers of convenience foods. And many thousands of square miles of primary forest are, with the support of organizations such as the World Bank, being cut and burned to provide land for the urban poor—land that often proves worthless two years after being clearcut. What are the repercussions of all this?

Tropical forests are important carbon dioxide consumers and are the pivotal point of the carbon, carbon dioxide, and water cycles. The metabolic processes within them help maintain our atmospheric composition. Light refraction from the tropical belt affects weather conditions which in turn mediate the climatic pattern of the entire planet. It is certain that tropical deforestation will alter the global climate. Already the burning associated with deforestation is producing so much carbon dioxide as to help contribute to an enhanced global greenhouse effect which is measurably warming our planet. In fact, referring to the carbon dioxide problem alone, Norman Myers (1984) concluded, "We can envisage no greater environmental upheaval, short of nuclear war, than climatic disruptions of this type and on this scale."

Tropical forests are golden geese in camouflage. One of the most important functions these forests serve, even small ones, is that of a watershed. Repatedly and tragically for people in India, Southeast Asia, Malaysia, Indonesia, the Philippines, South America, and Africa who depend on the natural cycles of creeks and rivers to raise their crops, removal of tropical forests has destroyed the land's ability to absorb, store, recycle, and moderately release rainwater into watercourses during the dry seasons. Instead rivers climb to record peaks and cause damaging floods and massive erosion of soil up to 6,000 times greater than normal! Then, during the dry seasons, once perennial creeks dry up completely. This is happening in textbook fashion in southwest Uganda and is part of the pressure forcing Bakiga into Toro District. It is also known now that removal of tropical forests actually changes the local climate: rainfall lessens and what does fall is more concentrated and destructive.

Beyond environmental services, the biological and genetic diversity of tropical forests is phenomenal, far greater than any other ecosystem. In *The Primary Source*, Norman Myers' celebration of the magnificence of tropical forests and outcry at their accelerating destruction, he notes that at least 40 percent of *all* species of plants and animals on the planet live in tropical forests covering less than 7 percent of the land. Edward O. Wilson, Harvard entomologist and author of the landmark book *Sociobiology, the New Synthesis,* considers this an underestimate and in 1985 warned, "we do not know, even to the nearest order of magnitude, how many species there are in the world." But he added, "recent studies in rain forests and other major habitats indicate the presence of as many as 30 million kinds of insects alone." Catherine Caufield cited a good example of the vast contrast in species diversity between tropical and temperate forests: "Mt. Makiliang, a forested volcano in the Philippines, has more woody plant species than the entire United States." According to Roger Lewin (1986b), in Borneo Peter Ashford, a Harvard botanist, found 700 species of trees, as many as in all of North America, in ten small plots adding up to one-tenth of a square kilometer! Lewin (1986a) also reported Terry Erwin, a Smithsonian entomologist, having identified 41,000 species of insects in a one-hectare plot, 1 percent of a square kilometer, in the Peruvian Amazon! Many botanists consider the tropical rain forests as the source for all modern temperate plants. P. W. Richards, one of the pioneer rain forest botanists and plant ecologists, described the rain forest as "the home par excellence of the broad-leafed evergreen tree, the plant from which all or most other flowering plants seem to have derived." But the tropical forest ecosystem is now vanishing on a worldwide scale. Just how bad is it?

Prior to the advent of agriculture 10,000 years ago, tropical forests covered 770,000 square miles; today less than half remains, though as late as 1967 about 60 percent remained. The official report of the Eighth World Forestry Congress (1978) stated, "On present knowledge the tropical rain forests are being destroyed at a rate of about 30 hectares a minute (0.12 square miles). The rate of destruction is accelerating. If it continues, these forests may cease to exist as usable forest in 40 to 50 years." This equates to an annual loss of 100,000 square miles. But estimates for rates of destruction vary. F.A.O. estimated an annual loss of 15,000 to 23,000 square miles and predicted total destruction in 60 years. But their more optimistic estimate did not consider any increase in the human population. The annual figure cited by the World Resources Institute is 30,000 square miles (an area as large as the state of Maine). Based on accurate remote sensing data, Norman Myers estimated the *irrevocable* annual loss of tropical forest at 35,000 square miles. The 1980 estimate by the United States National Academy of Sciences is even more sobering: 78,000 square miles of tropical forest were being destroyed or seriously degraded annually worldwide. Paul and Ann Ehrlich, veteran biologists and early conservationists, predicted that within the next twenty to seventy years there will be, "no tropical forests at all . . . except for a few scattered reserves and small patches on hillsides too steep to be put to any other use." In many tropical nations forests are not expected to last beyond the year 2000. Most tropical biologists agree, based on present trends, that *all* tropical forests will be gone by 2100.

In Africa about 8,000 square miles of tropical forest disappear every year, and, again, more than half of the original forests are now gone. What remain, mostly in Zaire, constitute 18 percent of the world's surviving tropical forests. Though astounding, and accelerating, this rate of destruction is lower than in Southeast Asia and Indonesia. Because of climatological disruptions during the Pleistocene, most tropical forests of today range from 4,000 to 12,000 years old as continuous habitat types. But refuge areas, including the small forests of Uganda and other sections of Central Africa, plus the Indo-Malay realm, are much older, some continuous from the age of the dinosaurs, and have served as sources of genetic diversity for younger forests. In conservation programs these forests should receive priority, although as early as 1952, P. W. Richards concluded, "It is therefore likely that the destruction of the tropical rain forest during the past 100 years has changed fundamentally the future course of plant evolution and closed many avenues of evolutionary development." If tropical forests are allowed to vanish as current destruction indicates they will, Norman Myers warned in 1984, "the planetary eco-system will need between ten and twenty million years to restore the damage done to the fabric of life."

Sadly all the drastic repercussions already mentioned that would result from the demise of the tropical forests are minor compared with the most critical consequence: the loss of species diversity. Without the tropical forest ecosystem countless millions of species of plants and animals who are obligate denizens of the forest will perish utterly, not just as individuals but as species. And, as Jon Roush of the Nature Conservancy pointed out, these wholesale extinctions are presently in the works, not merely hypothetical. The current rate of species extinction is at least 400 times higher than normal (i.e. pre–1600 A.D.; although evidence presented in Paul Martin and Richard Klein's *Quaternary Extinctions* suggests that pre-agricultural man also may have been a potent factor in species extinctions).

If current trends continue, approximately 20 percent of the species on earth in 1982 are expected to be extinct by the year 2000. Roger Lewin (1986a) quoted David Raup of the University of Chicago, who pointed out that the expected level of extinctions by 2100, 50 percent of all the species on earth today, would compare closely with the catastrophic extinctions 65 million years ago at the end of the Cretaceous, when the most advanced dinosaurs, along with most of this planet's other species, disappeared. This loss of species diversity leads to increasingly unstable ecosystems, strips variety in life from the planet, and loses irretrievably valuable but unexplored resources. Less than 5 percent of the world's plant species, for instance, have been investigated for pharmaceutical purposes. One ironical twist is that no one even knows the full magnitude of what is being lost; tropical forest biology is in its infancy. Lewin (1986b) reported Thomas E. Lovejoy of the World Wildlife Fund admitting that perhaps only one in ten or one in twenty species in the tropics is even known to science, then Lovejoy asked how intelligent decisions on conservation and economic development can be made in the face of such ignorance. Donald Perry, a pioneer tropical ecologist, likened the forest canopy, that part of the forest most crowded with life but virtually unexplored, as a whole new continent of life remaining to be discovered.

Economic considerations aside, does *anyone* have the right to commit the ecological devastation, biological genocide, and esthetic vandalism attending the destruction of tropical forests? When one considers that this exploitation, at rates precluding a renewable resource cycle, benefits only a tiny fraction of one or two generations of people, but consigns all future generations to a fate of ecological instability and environmental poverty, the answer is obviously no. Not only does the global community need a reevaluation of its economic priorities, humanity requires a shift in attitude toward the planet that nurtures it. Aldo Leopold, the pioneer wildlife biologist, stressed this concept most cogently: "Examine each question in terms of what is ethically and esthetically right, as well as what is economically expedient. A thing is right when it tends to preserve the integrity, stability, and beauty of the biotic community. It is wrong when it tends otherwise." Leopold termed this perspective the "Land Ethic," an attitude essential for human civilization to continue enjoying the greatest benefits offered by Mother Earth.

So who is responsible? Michael Robinson of the U.S. National Zoo (quoted by Lewin [1986b]) warned against the "enlightenment fallacy," the notion or claim that by educating the Third World they will come to their senses and halt the destruction of their tropical forests. "The problems are not due to ignorance and stupidity," Robinson said. "The problems of the Third World derive from the poverty of the poor and the greed of the rich." When all is said and done, forests are cut to be sold for cash. And when they are cut at rates precluding self-renewal they are being harvested for more cash quicker. And when they are not replanted or cut in ways that allow self-regeneration, it is because some of that cash would slip out of the hands of the timber concessionaires and be "wasted" by dumping it back into a forest that would not be harvestable during their lifetimes. And the reason that this system persists is because so many Third World countries have little or nothing else to sell for foreign exchange on the international market, a buyer's market.

If a view of earth from space were easily accessible to all of us and we could speed up a little what we saw, as in a time lapse effect, it would appear that the tropical forests were being gobbled up before our eyes. We could not help but ask ourselves what manner of voracious creature it must be to devour entire forests so quickly. What *is* the creature that ate the forests?

The Creature that Ate the Forest

You . . . and me. The human race. Some of us are doing it, others are supporting them, perhaps unknowingly, and most of the rest of us do not care enough to go out of our way to stop it.

Who is fighting this destruction? The total money spent on tropical biology in 1986 was approximately $50 million (less than the cost of a single MX missile), $30 million of which was from the U.S.A. For perspective on the adequacy of this amount, E. O. Wilson (quoted in Lewin [1986a]) admitted that the estimated cost for a five-year budget to make an impact on tropical deforestation is $8 billion, an amount more than thirty times greater on an annual basis. One worthy project alone, one proposed by George M. Woodwell of Woods Hole Research Center and his colleagues, Richard A. Houghton and Thomas A. Stone (described in a *Scientific American* staff article, "Seeing the Forest") to measure global rates of deforestation extremely accurately by painstaking comparisons of available Landsat imagery, would cost $5–10 million. Curbing the damage requires money in quantities available to neither ecologists nor conservationists. Hence the growth and evolution of conservation organizations.

Although I eventually came to recognize the global problem as the greatest single challenge in the history of biology, at this point I felt my greatest concern over threats to the chimpanzees of Kibale. In the north the Forestry Department was cutting timber and spraying arboricides on undesirable species, including figs. Although the central and southern blocks of the forest currently were safe, the same could eventually happen to them. Other real threats included adverse changes in the legal status of the forest, poaching, and illegal encroachment. (In 1982 Van Orsdol found that the Bakiga had settled illegally on 37 square miles within Kibale Forest [16 square miles of that in actual forest habitat], 17 percent of the entire reserve! By 1984 the government had removed about half these people to alternative land.) Some species in Kibale already had been extirpated: the hippo, lion, and probably leopard, and all other big game now were on shaky ground because poachers harvest wild animals at far higher rates than these populations replenish themselves. If unresisted, poachers would wreak absolute havoc. And, no matter how I looked at it, the fate of the apes of Ngogo was inseparable from that of the forest itself. The main challenge was to find a way to safeguard the entire forest. The only solution that came to mind (other than buying it) was to gain national park status for the central and southern blocks.

A simple enough solution, but national parks in the Third World must pay for their own keep, and tourism in Uganda was at its nadir. Economics here counted as much as anywhere else. Unless Uganda gained economically by gazetting Kibale Forest National Park, it would remain a forest reserve. Kibale needed publicity to get on the touristic map. I started thinking about how to achieve this, and this book was one result. But although my second question, "How long can Kibale Forest be expected to survive?" continued to haunt me, I had come here seeking the answers to other questions, many of which had continued to elude me.

$$\text{------ } 6 \text{ ------}$$

A Day's Work

N ight had fallen over the forest. Tiny phosphorescent flickers betrayed the mating rituals of drifting fireflies, and huge bats created erratic blotches against the tropical stars. Mist rose from deep forest valleys in ghostly white. A spiral horned bushbuck stepped delicately from the dense tangle at the edge of the forest and cautiously continued onto the sward at Ngogo Camp. Through this scene of deceptive quiet advanced an army whose soldiers knew no mercy; they would kill and devour any creature in their path who could not escape.

Millions of jointed legs clawed and scrambled through, around, and over blades of elephant grass. Antennae wriggled incessantly to test the air for prey. Uncounted nervous systems were tuned to zero in on abrupt vibrations; the insect army would swerve toward and swarm over anything causing a disturbance, such as a struggling body. Because the soldiers were nearly blind, trails of pheromones, communicatory hormones, were blazed as the outriders and vanguard advanced. Leading workers dropped back to be replaced by new leaders as the main body followed implacably. A faint rustle was the only warning of its approach.

The army ants filtered from the tall grass onto the sward. Progress increased because the vanguard no longer made false starts up pillars of grass to swarm over sleeping insects, birds, and lizards. Looming ahead was a dark structure of dried mud fissured with scores of cracks leading into the interior. Hundreds of workers, accompanied by soldiers half an inch long and armed with outsized mandibles, arrived at the fissures, entered them, and passed within. Hot on their heels swarmed an army of a million or more.

Inside I suspended a dented and cracked hurricane lantern behind my bed of elephant grass, then I slipped under my sleeping bag and opened a book by James Michener. The forces of Islam were undermining a Crusader's stout castle, while the doomed Christians within waited with the fatalistic calm of martyrs certain they would go fighting to the last soul. I wondered how Michener would handle the inevitable massacre.

Abruptly Simon began swearing on the opposite side of a partition of elephant grass. "*Siafu* (ants)" was the key word in his diatribe.

Secretly elated that the little devils had come through the walls on his side of the hut instead of mine, I slipped my boots on, flicked away the few ants, quickly gathered my fatigues, foam mattress, and sleeping bag under one arm, and rushed into the night. The bushbuck barked a startled warning and crashed into the grass.

My destination was Old Rover. It had a front seat too short for me and a bed nearly two meters long. The ground was too wet; besides, *siafu* make lousy bed mates. I looked into the bed and saw that it was flooded with rainwater behind the cab. I threw my mattress, bag, and clothes on top of the doomed sheet metal canopy over the bed, then climbed up to laugh at the ants; this time not one had nipped me. Sheet metal buckled and popped like firecrackers under my weight. Simon silently crawled into the cab.

Now supine, I gazed at the mist obscuring the stars. I hoped the rain would hold off until daylight. I was sharply aware of the flimsy metal sloping away from each side of my backbone. A slight move to either side and I would roll off and fall seven feet to the ground in the dark while trapped in my sleeping bag. I carefully adjusted my position and set off more firecrackers. Despite everything, sleep came quickly.

Plop. . . . Plop. . . . Plop. . . . Heavy drops thudded on my bag. I looked into the sky and more rain splattered on my face. I wondered if the ants were still swarming in the hut. My watch said 2:30 A.M. It was worth a look. I cautiously slipped out of my bag, creating a din that seemed deafening, then jumped into the darkness to land with a thud on the sloping sward. I made a beeline through the hut to my flashlight and used it for a hasty inspection. No ants in my corner, but the rodent skulls on the elephant grass dining table were crawling with them. I unrolled my bed and slipped into it. Already the rain was pounding against the iron roof.

Army ants crawled through my dreams. Four meters away they crawled over the skulls and last night's leftover beans; their combined strength lifted the lid off the pot with an audible clatter. This was a windfall for the ants. My gamble that they would not backtrack paid off.

The morning was gray and wet and offered little incentive to get out of my bag, but I was ravenous. Just after dawn, Simon dragged himself in and silently pointed toward the reddish-brown mat of *siafu* smothering the pot of beans. Our food supplies were ebbing. I ate some cold potatoes baked in our cooking fire, almost savoring the residue of ash, stuffed several more of them into my pack along with some roasted groundnuts (peanuts), emptied half a thermos of tea, then plodded into the forest.

Siafu usurped me in the mud hut on only ten nights during two years, but because I lived there alone most of that time, I could not count on someone else providing advance warning. Twice my first inkling consisted of their bites. When one ant bites or stings it releases a pheromone that communicates to other ants that it is time to bite. They all attack like well-programmed machines and do not stop until they are removed from one's body, and sometimes not even then. Mother Nature was in a sour mood when she conceived of army ants.

Eventually I conditioned myself to awaken at the rustle of their approach in time to escape. Colonies of African army ants may number more than 20 million individuals, according to Edward O. Wilson (1971), collectively weighing forty pounds—frightening statistics. As a child I had been fascinated by Carl Stephansen's "Leiningen Versus the Ants," the classic tale of a plantation owner battling to defend his crops against a swarm of army ants. It had convinced me they were the ultimate doom. Now, even though I knew they traveled so slowly that they can easily be avoided, some of that fascination remained. The trick is not to be caught unawares or immobile. More than one man of Toro District has passed out in an alcoholic stupor to be eaten alive by migrating *siafu*.

But *siafu* were only one down side of life in the mud hut at Ngogo. The hut itself, as mentioned in Chapter 2, consisted of a framework twenty feet square of poles of *Diospyros*, a semi-straight-growing tree, which then was laced with bundles of elephant grass tied to the framework with small lianas (vines). Yongili and Clovis had packed the basket-like walls with mud. After drying, clumps of the walls chipped, leaving holes and a latticework of cracks. Eventually a second, then third coat of mud tidied them. Eliminating the fissures was important not only to reduce wind and rain blowing through them; it also cut down on unwanted traffic, including snakes who oozed in during the night.

Although by now we finally had nailed in wooden shutters and a wooden door, for the first three months we lived in the hut, eight irregular holes served as drafty and sometimes wet windows, and a one-by-two-and-a-half–yard opening was the doorway. A buffalo could have walked through it. More important, so could a spotted hyena. In some parts of Africa hyenas have been known to bite off a sleeping person's face, hands, or genitals. Most likely this habit resulted from hyenas finding unburied corpses outside villages. Happily, hyenas at Ngogo turned out to be more fastidious.

The hut was an artificial cave and a scarce resource that attracted cave dwellers immediately. Our most stable housemate was a male *Agama* lizard, bright metallic blue and yellow with a turquoise tail. He stalked and ate many of the insect inhabitants, especially the crickets who joined for nightly sing-alongs in the wall cracks. I measured the heavy *Agama* as he slept clutching the wall, fourteen inches. Because he must have looked like a monster out of a nightmare to the insects who let him too near, I named him Godzilla.

A Day's Work

Godzilla had a mate, a gray svelte version of himself about two inches shorter. She restricted her foraging to the south side of the house to steer clear of Godzilla, who would not allow another male *Agama* near the place and was not completely captivated by her charms either, especially when it came to the question of which of them would eat. She usually slept anchored to the wall two feet from my bed, where her body merged with the rough brown mud. Eventually I found myself saying good night to her.

I installed an empty oil drum Struhsaker had acquired under the end of the rain gutter draining the roof to reduce the loads of water that had to be hauled from the creek in the forest. Not long after I installed it, Yongili, back from his leave, exclaimed, "Your *Agama* is in the drum, already dead!" I rushed out to see Godzilla's mate belly up and spread-eagled in the water with her head submerged. No way of telling how long she had been in there. Much to the disgust of Yongili, who considers lizards taboo and will not touch a chameleon for any incentive, I fished her out by the base of the tail and placed her on the ground. For a while not much happened. Then she blinked, looked around, and scurried up the wall. Uncharacteristically, Yongili was at a loss for words.

Between the top of the wall and the lower beams of the roof was a six-inch gap, a highway for lizards and other creatures able to scale the walls. In the evening small bats lost to the thrill of the chase cruised in hot pursuit of insects seeking sanctuary within. Having discovered the place, a few bats suspended themselves from a beam to sleep out the day, but most did not stay because the corrugated iron roof gathered a lot of heat. On a reverse schedule, swallows occasionally used the same beams for night sleeping perches.

The least desired visitors came at night. When the crickets in the cracks sang and I tried to sleep despite them, marauders scaled the walls, scurried across the beams, and descended to gorge on our food stores, which for lack of protected containers lay in baskets woven from papyrus. Groundnuts drove them crazy. Ngogo rats swarmed to sit in a basket, munch, then excrete back onto the pile. Where these were exhausted, they gnawed any other container of roasted groundnuts. They drilled through plastic containers even if they now were empty but once held groundnuts. One gnawed through my pack, then through my bag of groundnuts. After hitting this glory hole the rat carried each nut to the privacy of my left boot, in which it sequestered itself to savor its booty away from the prying eyes and noses of its comrades. This was the last straw.

I set three break-back traps baited with groundnuts. I awakened at the report of each trap to reset it and toss the corpses out the window for the African civet cat who snapped them up during his noctural patrols of the sward. I carried a flashlight and spear because of my frequent opportunities to spear a trapwise rodent. My overzealous spear thrusts sometimes knocked holes in the walls that caused Yongili to wring his hands.

The worst assault was launched during the current month, May, on my first night back from Kampala, after having dropped off Simon at Entebbe International for his return to England. The hut was overrun by an occupation force. They knocked pots over, scurried through specimens and books, and ate almost everything, then nosed into nooks, crannies, and myself in search of more. This was a battle for survival. That night I trapped three and speared four more on the run, then got a dozen more on the next three nights. Eventually I relied more on the traps than the spear because the mud walls dulled the blade, and I was becoming groggy from lack of sleep. But their tenacity and reproductive proclivities were so prodigious even in the face of my defense that they eventually forced me to store much of my food in Old Rover. But *still* they came.

Despite its leaky roof, fissured walls, chiggers on the floor, invasions of army ants, insatiable rats, buzzing crickets, snakes, bats, cane rats, birds, lizards, insects, and other vagrant traffic, this mud hut became both home and my base of operations. My life revolved around a cycle of nights in the hut and days like the following, an actual day taken from my notes, in the forest.

With the first glimmer of equatorial dawn at about 6:00 A.M. I sat up in my bunk, slipped on my shirt, and began twenty minutes of meditation. My mind was haunted by disappearing wisps of unusually vivid dreams that made my night seem as active as day, though not quite as explicable. Then I slipped on my trousers, socks, boots, and gaiters; they were still damp from the previous day and felt anything but comfortable until my body warmed them. I poured a pint of strong Ugandan tea from my stainless steel thermos (unbreakable) and dumped in enough coarse sugar to make a dentist switch to driving taxis. I gulped down some vitamins and chased them with tea. As quickly as I could, I gobbled a half dozen or so small bananas, then a handful of groundnuts, then more bananas. The forest was now clearly visible through the window and I realized that I should have been out there already. I slipped several more bananas into my pack, ascertained that my groundnut container was full, hurriedly brushed my teeth, then pocketed a few *Pterygota* leaves.

While crossing the sward en route to the latrine hut, I scanned upward across the grassy hillside. Already I had seen waterbuck, bushbuck, buffalo, and elephant grazing within a couple of hundred yards of camp in the mornings. But this morning the hillside was empty.

The forest was wet and the trail muddy. I kept a sharp lookout for army ants, and beautiful but lethal snakes. I tried to keep each step quiet. Chimpanzees could have been anywhere and I wanted to see them before they saw me. Some individuals still fled when they spotted me, so I preferred to see them first, identify them, and watch them by stealth when necessary. Usually I had only a rough idea where the apes were on a particular day, and on about half my days in the forest I failed to find any despite a full day of searching. Any clumsy noise now could mean a fruitless day.

A troop of red colobus and a nearby harem of redtails filled the trees ahead of me. I had to pass them. Some were likely to notice me and give an alarm call that would set off a chain reaction of other alarms and a blizzard of monkeys leaping and crashing in unnecessary flight through the foliage.

I slowed down and proceeded cautiously, taking advantage of natural cover. I slipped safely under the noses of the colobus, but a female redtail spotted me and gave a half-hearted alarm chirp, more in surprise than alarm; the monkeys near camp were mostly habituated to people. A few red colobus craned their necks to discover what was disturbing the redtail. One saw me and gave another alarm call of slightly higher intensity, but by now I was past them and the clamor died down quickly.

I continued down the trail. A red duiker (a small antelope) leapt off it, swished through the understory, and gave a wheezy bark after a second or two. The sun appeared and I stopped to admire the view across a swampy creek course, the only lengthy view to be had in the forest. All the while I listened for pant-hoots or other sounds of chimpanzees. Unaccountably, a song by Neil Young, "Cowgirl in the Sand," started playing in my head and would not stop.

A narrow brown ribbon crossed the trail ahead. I stopped to watch it. Like an animated cartoon version of rush hour freeway traffic, thousands of army ants jostled one another in a frantic hegira from one bivouac to another. Workers rushed along with developing larvae tucked up under their abdomens, while soldiers trod gingerly, as if hopelessly clumsy and unable to avoid stepping upon the stream of workers. Other soldiers stood on either side of the column on a moat-like rise of granulated mud that seemingly contained the flow of ants. These soldiers faced outward with their huge mandibles as a first line of defense. A day from now, after an uninterrupted flow of a million or more ants, the rear guard would be traveling through a completely enclosed tunnel of mud constructed by the earlier migrants. Meanwhile they would have consumed as many as 100,000 arthropods. This migratory phase of army ants may continue for two weeks, according to T. C. Schneirla, a tropical entomologist, until the larvae spin their cocoons and become pupae. When this happens the swarm will begin its "statary" (stationary) phase and the queen will lay up to 100,000 eggs in a week. After a few weeks the pupae will metamorphose into little callow workers in synchrony with the hatching of the new eggs, and the colony will become nomadic again.

I paused above this roar of tiny machine-like lives and felt like a vengeful demigod. The temptation was strong to stomp my boot onto that orderly confusion to repay their kind for the indignities to which they had subjected me. But my boot would have taught them nothing; it neither would have prevented nor exacerbated their future attacks on me. These sterile sisters could not understand that, in this life, communism benefits only the ruling class, in this case, the queen. Actually, they were acting strictly in their own best reproductive interests. Remember inclusive fitness? The sociobiological explanation for why these workers were more willing than the most fanatical kamikaze pilot to die for their colony, a willingness genetically hard-wired into them, was that their only hope of reproducing the genes they carried lay in helping their mother, the queen, to produce even more sisters, some of whom would become queens. I stepped over the migrating column and the tune resumed, "Hello, ruby in the dust . . ."

A flicker of movement caught my eye. A red-legged sun squirrel darted along a crooked limb in a *Pseudospondias* tree growing on the edge of the swamp. I focused my field glasses. The squirrel paused, plucked a green fruit the size of an olive, stripped away and discarded the pulp, and tucked the seed in its cheek pouch, possibly for storage later in a hidden cache. I wrote this in my notes under the rubric, "Opportunistic observations of mammals and birds." Chimpanzees eat the fruit and seed of *Pseudospondias*; this squirrel was a competitor.

As if from some hidden epicenter in the forest, sonorous frenetic drumbeats reverberated through the earth itself. I knew the source was Bigodi, seven miles as the gray parrot flies, where people were celebrating something. Why and how the forest floor resonated like a sounding board had ceased to perplex me by now, but the drum beats still sent an involuntary wave of apprehension up my ribcage, as if they boded evil. They also dredged up the well-used line from B-grade adventure movies, "The natives are restless." I continued to scan the crowns of trees. In the back of my mind I could not help but wonder if in fact they *were* restless.

About a third of my time in the forest was spent on vigils at large fruit trees, the most certain way of contacting the chimps. No vigil today, though; this was a lean season and I knew of no big trees especially attractive to the apes. Contrary to what many people assume, the tropical forest, at least this one and the many others I have seen, is not festooned with gorgeous orchids or scattered with colorful blossoms. The forest is a study in green; it contains nearly every imaginable hue. Ngogo was home to at least one hundred species of trees, not to mention hundreds more species of epiphytes and understory plants, and these plants did bloom, often with blossoms of complex and delicate beauty. But for unknown reasons, sometimes years on end passed between one flowering and the next. Normally the forest was green and more green. As I scanned the canopy, color was one of the main stimuli triggering my interest. Chimpanzees snacked on the blossoms of many species, several that I had yet to witness. So far, though, nothing had caught my eye. I was relying on woods lore and intuition.

But I did know that chimpanzees were in the study area. A couple of hours after midnight this morning I listened to a chimp in the north who had repeatedly screamed and pant-hooted, and then was answered by one in the south as much as a mile distant. Chimpanzees often communicated thus at night. I imagined them snug in their leafy nests high above the forest floor with a case of insomnia. Suddenly one wonders if so-and-so is nearby and pant-hoots in hopes of eliciting a response. Sometimes these nocturnal calls set off a chain reaction of pant-hooting choruses by apes at three or four locations. The resulting bedlam was difficult to keep track of in the wee hours of the morning when, despite my having conditioned myself to awaken, listen, determine probable sites of origin, then write those data in my notes under "Pant-Hooting at Ngogo," fatigue and strange dreams clutched at me with soporific fingers.

I continued south. By now the approximately fifty miles (Yongili had added another half dozen I needed) of cut trail in the study area had become as familiar as a lover's face. But the forest was not stable. Giant trees died and collapsed thunderously, sometimes dragging a retinue of smaller healthy trees with them across a trail. I cursed the wind that toppled them as I climbed up and over. The tune stayed in my head, "Hello woman of my dreams . . ."

Through a tiny sunlit meadow I approached a miniature creek, then halted in my tracks. Its visual cross section could not have been much larger than a roll of 35mm film, but atavistic neurons in my brain magnified it to the biggest bit of information in the neighborhood. Eyes black as tar stared at me from a setting of polished lime-green scales. The forest mamba was coiled in the creek. Perhaps vibrations from my tread had alerted it to raise its head above the water; its yellow belly and green back blended with the lush grass where the path crossed the creek. Unless I detoured widely, I would not feel happy detouring at all because of the bad visibility and the chance that I might run into what I was trying to avoid. So I edged closer in a game of bluff. No poisonous snake considers an adult human a potential meal, but we are threats. This one had plenty of room to retreat and it did. When it had gone about fifteen feet, I jumped the creek.

On the trail ahead fresh chimpanzee dung was already attracting dung beetles. Scarabs (dung beetles) actually cruised along the trails about a foot or two above the ground, presumably scanning for dung. These two had shaped it into a ball containing all but the bulkiest components. The pair of black scarabs, each hardly even a half inch long, were rolling a ball that dwarfed them. The female ascended the high side and, by clutching it and leaning outward, her tugging unbalanced it and rolled it in her direction. Simultaneously the second scarab cooperated by standing on its head on the opposite side, bracing its hindmost feet against it, and shoving upward to rotate it toward the female. They rolled the ball over rough terrain littered with twigs and other irregularities higher than they were.

The pair labored for almost two minutes to propel the ball eighteen inches to the base of a tree. The male dug until he disappeared into the detritus with the ball on his back, like Atlas holding the earth. His partner guided the ball downward. Both scarabs and dung were buried after ten minutes.

Dung beetles are ubiquitous across Africa. Entomologists Bernd Heinrich and George Bartholomew reported that the continent has more than 2,000 species ranging in size from tiny miniatures to huge ones weighing more than the smallest mammals and birds. They concluded that the scarabs' rapid and competitive internment of dung for feeding and egg-laying tends to fertilize the soil, retard the spread of some diseases and parasites, and reduce the reproduction of dung-breeding flies. In Kibale Forest they kept the trails clean and occasionally planted seeds for a future forest.

About two miles from camp I approached an open glade interspersed with many tall, straight *Diospyros* and *Uvariopsis* with interlacing crowns. The scant understory gave the place the unusual effect of a cathedral. As I paused to contemplate this, a limb in a *Pterygota* beyond the glade vibrated. Pay dirt.

I moved silently into the *Pterygota* grove. An adult male, Shemp, and a small subadult male whom I did not recognize were munching on the turgid pale wings of immature seeds. Fragments of seed pods and discarded seeds rained on the ground beneath them. They foraged for twenty minutes as I spied on them, then Shemp built a nest.

In less than a minute he pulled five or six leafy boughs beneath him, almost without rising from the horizontal forked limb on which he was sitting. He settled back into the nest at peace with the world. He leaned back further, let his head loll to the left, and found himself staring directly at me thirty yards away.

He sat up straight as a ramrod and glued his eyes on me. He parted his lips slightly, and I anticipated him vocalizing a clear hoo that would have alerted his partner. Instead of hoos, though, Shemp emitted little peeps that seemed so ridiculous and incongruous coming from his muscular form that I had to turn my head to stifle my laughter so as not to startle or offend him any more than I already had. After a minute or so of peeps, he rushed out of his nest, kicked a dead limb to send it crashing out of the tree, then moved to an open area and stopped. We both scratched ourselves and covertly glanced at one another. If I had not surprised him so completely when he least expected it, I think Shemp would not have been so upset with me. It was as if he was embarrassed at having been caught unawares and was trying to obliterate it through bravado. He finally climbed higher and disappeared. For a while he was quiet, then he resumed foraging on seed wings.

From at least a hundred yards northeast a chimpanzee screamed repeatedly. Neither male vocalized in response. Instead they alternated foraging with lying around for another hour. They neither groomed one another nor interacted directly at all, but somehow they seemed to emanate a subtle solidarity. About two hours after I joined them, they simultaneously descended to the ground, glanced at me nearby, then walked quickly to the northeast, vaguely in the direction of the apes who screamed earlier. I followed more slowly, wishing these two trusted me enough not to bolt when I tried to keep up with them.

Again I sneaked along as if hunting creatures possessing uncanny powers of perception. I crossed a creek and stepped around deep pothole tracks left recently by elephants, but noted that knuckle marks of chimpanzees were absent. Already I was off their trail. I crossed a second creek, headed up a muddy slope, and scanned the trees for chimpanzees. Two yards ahead another forest mamba swished its jewel-like train of green and black scales across the path—death with a lemon yellow belly. Again I was glad they do not eat people; it might have had me. If it had bitten me, days or weeks might have passed before any who searched might have found me. These were the most common snake in the forest. I wondered if they were as dangerous as their reputation claimed; I never did find out.

Eons of evolution have produced birds adapted to hundreds of ecological niches in the forest: fruit-eating hornbills, the male of which walls up the female and their clutch and feeds her through a slit, kingfishers that rarely go near water, weaverbirds that construct symmetrical nests rivalling a human artifact, pigeons living on fruit, and eagles specializing on serpents. Curiosity spurred me to identify the birds I saw, but because Africa was home to many unfamiliar families (hoopoes, turacos, bulbuls, coucals, sunbirds, bee-eaters, wagtails), I often found it a challenge. By the time my notes on descriptions and behavior identified a species, I usually felt intimate with it. Occasionally I even carried ailing birds to camp to nurse them back to health. In a rare instance of natural justice, some species assisted my project by advertising the presence of fruit, or even their primate competitors. I listened for bird calls now, but these helpers seemed nowhere near today's route.

Only half a mile from camp I had to recross Kanyanchu Creek, the main watercourse draining Ngogo. As I was backing up to get a run at leaping the swollen coffee-colored creek, I heard a faint plopping from downstream. I froze. Sleek and powerful, a Congo clawless otter appeared swimming upstream. At nearly forty pounds and nearly five feet long, it barely fit in the creek.

It stopped in midstream and turned its streamlined bewhiskered face toward me. It blinked and squinted as if the silty water had hazed its vision, gazed at me for a couple of seconds, then submerged to be drawn downstream a short distance. A brief encounter. I remained still on the chance I might see it again. In a few seconds the otter swam back to two yards from me, paused, squinted, and studied me again. I found it hard to suppress a smile, but I did not move otherwise for fear of startling it. Again it submerged and drifted away only to return for a third inspection.

As it drifted downstream a third time I remembered the camera (now repaired in Nairobi thanks to Lysa) in my pack. Would it return for a fourth look? Did any photograph exist of a Congo clawless otter in the wild? They are very rare; and this was the only one I had seen. Quickly I unslung my pack, knelt, groped for the camera, and, as I grasped it, the otter returned to scrutinize me with large eyes set in a face seemingly designed to express fun. I froze again, but my shift in position must have seemed suspicious. It quickly submerged, drifted downstream, then abandoned the creek to detour overland through the African ginger. I caught a last glimpse of it as it bounded across the trail with a short-legged, hump-backed gait, presumably to return to the creek upstream of me. I tried to anticipate its reentry point but missed and never saw it again. I later read in my field guide to mammals that little is known about them other than they forage in the understory and they are predominantly nocturnal.

The sun had already dropped behind the Mountains of the Moon as I emerged from the forest. Though gone for about ten hours I had covered only four miles. But it had been a reasonably good day.

Yongili had heated up water for a bath over a fire in the tiny, grass-walled kitchen hut. The large soot-covered pot near the fire contained my perpetual bean stew spiced with the maximum number of peppers short of suicide.

I entered the mud hut and dropped my gear next to my bed on my U.S. Army surplus rocket box, a waterproof steel box with welded hasps in which I stored my original data sheets and a few other things I hoped to keep securely dry. I returned to near the doorway and chinned myself from the horizontal pole I had nailed from the center support pole to the door lintel. Then I did one hundred sit-ups on my bed, nineteen more pull-ups, and headed out to the grass structure euphemistically termed a bath hut to dream of a real hot shower among the spiders and the mushrooms. I carried my pan of used bath water back to the hut and spread it over the rocky, sloping floor to keep down the dust, discourage chiggers, and help hard pack the floor when I swept it.

By dark my belly bulged with potatoes and avocados smothered in bean stew. Despite this, I ate another banana. Except for very limited supplies of sugar, flour, rice, salt, and sometimes powdered milk, virtually all my food was local produce. Amin's heavy-handed dabbling in Uganda's economy had dried up imports other than military hardware and luxuries for his circle of sycophants, so not even canned food or spices were available. I bought bananas, beans, groundnuts, potatoes, cabbage, carrots, avocados, onions, tomatoes, garlic, peppers, tea, oranges, papaya, pineapples, and eggplants when I could, plus occasional eggs, in the outdoor market in Fort Portal, in Bigodi, or, during hard times, anywhere. Chronic shortages of gasoline limited my four-hour round-trips to town to every three weeks, and the unpredictability of political developments also made infrequent trips to Fort Portal prudent. And, despite the occasional bananas, carrots, maize, and basil that the bushbuck left me in the garden, my cupboard often held only beans, potatoes, and tea.

After flossing my teeth I saved the thread for sewing repairs then walked outside to the rain gauge to brush my teeth and gaze across the forest.

The moon already was high in the east, and the filmy cloud of a myriad stars forming the Milky Way Galaxy was an arch across the black sky. Many miles to the south lightning flickered in silent sheets above the hills beyond Kibale Forest. The bushy-tailed civet bounded across the sward to vanish in the tall grass. The air was charged with glowing fireflies and stridulating crickets. From the southeast the eerie pant-hooting of a chimpanzee pierced the night as if a tangible entity and filled me with excitement, as if that ape alone could answer all my questions. The rising whoops of a distant hyena in the east told the forest the night was cover for a hungry hunter. Nights at Ngogo were a natural symphony.

I spat out Colgate toothpaste and hit the sack. Godzilla's mate appeared carved from slate. For an hour or so I read by the dim light of the old hurricane lantern. Here books were almost as important as mail, which sometimes arrived in Fort Portal but once a month, after Amin's censors were through with it. The complexity of the forest and the mysteries of the chimps dominated me, but I missed communication with others of the culture that spawned me. This was why mail was so important and books so much more real than in civilized lands dominated by commercial advertising that subjects one's senses to a constant bombardment of the highest-powered stimuli allowed by law.

I blew out the lamp and reached out to feel the smooth shaft of the spear leaning against the wall. The Southern Cross appeared low in the small patch of sky visible outside my window. There was so much I still did not know. I closed my eyes and again wondered where, exactly, the apes were and why they were there, and I wondered how far this quest to understand them would lead me.

7

To Be Groomed

With chin tilted to the sky and eyes half closed, Zira sat a dozen feet above the forest floor on a horizontal limb jutting from an enormous *Piptadeniastrum* tree emerging from the lip of a steep ravine. I heard faint popping as ancient Gray slowly smacked her lips several inches from Zira's throat and carefully parted sparse hairs to examine Zira's skin for foreign matter or dead skin. Because her right hand was partly crippled, Gray used her left primarily for fine manipulation of skin and hair.

Gray grasped Zira's head in both hands and bruskly yanked it downward so that now Zira was gazing at Gray's navel. Then she scrutinized the skin of Zira's beetling brow. Zira sat still for this operation and, when Gray twisted her head this way and that, she became as pliant as clay in the hands of a sculptor. Against the blue sky and flickering emerald leaves, purple-headed glossy starlings, white-headed wood hoopoes, great blue turacos, and crimson-winged black-billed turacos flashed across my view of the intently grooming apes. After two minutes of examining Zira's crown, Gray stopped abruptly, turned away, and lay prone, draping herself along the massive limb like a rag doll dangling an arm and leg down each side of it.

Zira blinked, then slowly leaned forward on her belly and elbows to stop with her chin only a few inches from Gray's shriveled perineum. With rapt concentration Zira examined this flaccid portion of Gray's anatomy that never again would catch the eye of a mature male, and she picked away bits of objectionable material. Gray was in bliss.

With my eyes fixed to field glasses to catch fine detail I stood about forty yards distant in a stand of saplings. I had been observing both females for the past three hours, during which time they had gorged on *exasperata* figs, built day nests and napped, eaten more figs, and for the last fifteen minutes had groomed one another. By now, July, neither female paid me much attention as long as I minded my manners. My field notes were beginning to bulge and I felt an inordinate fondness for my data, akin to what misers reputedly feel for their gold.

Behind me, maybe fifty yards up the gentle slope, a limb snapped.

My sense of well-being dissolved immediately. Early this morning the forest had resounded with primeval screams and roars that sent shivers up my spine. My trail cutters, Benedicto and Rhawire (replacing Yongili who was on leave again), were afraid to enter the forest. By spending an extra fifteen minutes reexplaining in Swahili which trails needed to be cut, I thought I had talked them into it. As they disappeared into the forest by one route, I entered by another. Soon nearby roars again vibrated through the greenery and Benedicto and Rhawire sprinted back up the trail to camp.

I swallowed the unpleasant wave rising in my ribcage and descended the west slope, then stopped at the narrow swamp. The trail was pock-marked with huge holes into which water was still draining. It looked like a giant had jammed a tree trunk into it about fifty times. From nearby in the southwest came loud roars that seemed to me excellent facsimiles of dinosaur diction. Almost immediately a horrific reply echoed from even closer in the northwest. My path lay between the authors of these unsettled sounds.

These vocalizations did anything but beckon me, but Ngogo was now in the middle of a very dry dry season and I knew that chimpanzees would visit the *exasperata*; it was currently the best tree I knew of for fruit at Ngogo. As if a private ritual performed only when Uganda seemed inimical, I asked myself again why I was here. The answer, as usual, was to study chimpanzees. Then I crossed the swamp. But I stopped frequently to scan the foliage and listen for the sounds of these mammoth creatures that could travel as silently as a whisper. Less than a mile beyond the swamp I found Gray and Zira and, later, Zane, a juvenile female who sometimes traveled with the two adults but who did not trust me. Then I congratulated myself for not having turned back at the swamp.

Another limb cracked.

Fear gripped me and, as some detached portion of my brain analyzed this novel sensation, I looked around for a good tree to climb. Nothing but small saplings; I was in the wrong place for this encounter. The scientist in me lifted my field glasses to determine the chimps' reactions to what now was an approaching stampede of elephants. I was disappointed to see them still immersed in grooming. Again I glanced around for a good tree to climb but nothing new had turned up. I could do nothing at this point that might not put me in a worse situation than I was in. I contemplated the fact that I was only one small entity in a large cluttered forest; maybe, if I stood perfectly still, they would not notice me and would just pass by me. In the back of my mind lurked the possibility that this might be my last moment; I regretted that I did not have more data. But my life failed to pass before my eyes; I was too busy watching for options.

Maybe poachers spooked them, maybe it was my scent in the air. As they ran toward me the elephants cracked limbs and snapped tall saplings off near the ground. Twenty yards to my right, one red-mud-coated pachyderm after another crashed past me, halted abruptly at the lip of the ravine yawning before them, then detoured southwest, away from me. Simultaneously, more elephants rushed past my left side to literally skid to a halt in the mud at the edge of the drop-off. These veered north, also away from me. In a few seconds a dozen elephants had pounded past me, the frail protoplasmic island in their stream. With shaking field glasses I again spied on Gray and Zira. They were *still* grooming and appeared unconcerned with the stampede! I felt betrayed.

Belatedly I returned to my usual vantage point for this *exasperata*. Though just as close, here more foliage obscured the view between the apes and me, but it was good enough, and it was safer. I looked for a good tree to climb should the need arise again. I felt stupid looking for an escape route now that the danger had passed, but I had no reason to believe that those had been the only elephants in the neighborhood, or that they would not return. A few yards away I found a perfect tree.

Now Zira was spread-eagled along the limb with her arms and legs dangling. Gray moved her grizzled, notch-eared head forward and leaned on her right elbow to groom between the sleek hairs of Zira's lower back with her lips and left hand. Both apes were totally engrossed. Today they would set the record for the longest session of mutual grooming I observed at Ngogo: two hours and three minutes, twenty-two minutes of which were lulls when one of them had gazed into the distance or had self-groomed rather than reciprocating when her turn came. Gray had groomed Zira twenty times (fifty-two minutes) and Zira had reciprocated nineteen times (forty-nine minutes) having examined most parts of each other's bodies at least once.

Social grooming was important to the apes of Ngogo. And it was predictable between preferred companions when tensions were at their lowest, such as after foraging. Chimpanzees, however, were blatantly more interested in being groomed than in grooming, and their sessions resembled dynamic but friendly competitions to be groomed the most. Typically, groomees presented one after another parts of their bodies in need of attention, until finally the groomer stopped and countered by presenting part of itself to be groomed. Roles then reversed, the new groomee was never at a loss in presenting additional areas requiring examination. The most impressive aspects of many sessions were the friendship, reassurance, and unqualified trust between partners whose personal space and normal body protocols broke down completely. The best way I have found to illustrate this is by using a description of a typical example I later observed of males grooming.

After wild pant-hooting and violent buttress drumming erupted from the base of a large *Ficus dawei*, R.P. climbed a slender access tree parallel to the twisted, fluted bole of the fig. His penis was erect, common when adult males arrived at large fruit trees, and he continued to pant-hoot, shriek, and grunt, pausing only long enough to stuff figs in his mouth, chew, and swallow them so fast that I wondered how he managed to avoid choking. This was food calling at its apex. Another male, Spots, climbed up to join him, followed by a juvenile male, Ashly.

As usual during foraging, all three males fed apart from one another for an hour. A tentative spearhead of redtail monkeys entered the crown but Ashly hurled himself out of the tree toward them, and they fled amid a chorus of high-pitched alarm chirps. A few minutes later he emerged from the foliage, climbed into the fig tree, and sat in a wide crotch. R.P., who had been sitting and gently scratching at the raw patch on his back, climbed toward him, lip-smacked in the special signal of a chimpanzee who advances toward a partner to initiate grooming, then presented his own back to be groomed.

Using both hands and his lips, Ashly carefully groomed around the wound. This was heaven for R.P. Eventually the younger male lifted his arm to present his armpit to R.P. for grooming, but R.P. deftly ignored this and turned to present his own back to be groomed again. R.P. finally reciprocated, then lay on his back with both legs spread, his huge-thumbed feet projecting skyward like those of a dead cockroach. Ashly meticulously groomed R.P.'s genitals with the care a mother devotes to removing a mote from the eye of her child. Finally, after having been groomed by Ashly for fourteen minutes compared to his ten minutes of reciprocation, R.P. climbed toward a clump of figs he had been eying.

While R.P. stuffed himself with figs for another half hour, his index finger repeatedly crept around to probe gently at the borders of his perpetual wound. He glanced at Spots, toward whom he had been edging for several minutes. He smacked his lips, planted himself behind the older male, and groomed his back. After four minutes of being groomed Spots climbed away to pick more figs as if he had not even noticed R.P. Disappointed, R.P. glanced around, then casually descended the access route. Spots noticed him and hurried toward the access tree with Ashly following.

The terminations of "doomed" sessions, like the one between R.P. and Spots, sometimes were delayed when an original groomer continued to regroom a reluctant partner, then stopped and waited again for reciprocation. These lulls, or "grooming standoffs," resulted from one partner desiring a continued session with an ape who showed no inclination to groom but did not mind *being* groomed. Sometimes patient apes eventually were groomed by these reluctant partners, seemingly only because of their persistence.

Hypotheses of two basic types have been advanced to explain the functions of mutual grooming: one is based on hygiene, the other on social facilitation. If one attacks them with Occam's razor, the most logical hypothesis is that apes who groomed one another reciprocally were engaged primarily in an exchange service: grooming another's inaccessible regions in trade for having their own groomed. The second popular hypothesis maintains that chimps (and especially other primates) groom one another as appeasement to ease social frictions.

Because I suspected the exchange-service hypothesis to be the most logical, I eventually combed my field notes for the hundreds of instances in which one chimpanzee groomed specific regions of another. Those parts of an ape which it can self-groom, the combined areas of the chest, abdomen, arms, hands, feet, and legs (excluding the back of the upper arms and thighs), were the target areas on only 4.3 percent of these instances, yet they account for 55 percent of the body surface! Conversely, those parts of a chimp's body which it could *not* see itself (the head, neck, back, perineum, backs of thighs and upper arms) plus awkward but feasible regions (shoulders and groin) received the other 95.7 percent of grooming. Thus the grooming-as-appeasement hypothesis made little sense in light of this evidence that social grooming is aimed almost exclusively at areas chimps cannot groom themselves. Chimpanzees who groomed mutually were doing one another favors on an exchange-service basis. (Independently Michael Hutchins and David Barash found that captive lion-tailed macaques groomed one another on the same basis.)

How important is this favor? The consequences in the wild of not being groomed are not well documented, although observations of solitary vervet monkeys by Struhsaker (1967) and of solitary baboons on the savanna by Sherwood Washburn and Irven DeVore revealed heavy infestations of ticks and other parasites not found on individuals who were being groomed. In Kibale ticks and botflies are common, and chimpanzees are also plagued by a species of louse detectable only by extremely close scrutiny.

In fact, when chimpanzee lice are exposed to light, as when hairs are parted to reveal them, they freeze and become very difficult to see unless one is very close (and not far-sighted). This makes it easier to understand why the apes give such exaggerated scrutiny to their grooming target.

The more I saw, the more convinced I became that to be groomed was the chimpanzee equivalent of receiving the check in the mail; it fostered a sense of well-being and, sometimes, near euphoria. This euphoria once broke down a major barrier for me.

It happened one morning at the same *dawei* as the example of R.P. above, only this time I had arrived to find Blondie with her infant son, Butch, and daughter, Bess, eating figs. Normally Blondie fled me, so I remained hidden while observing them. An hour later Owl joined them. Seven-year old Bess climbed to sit behind Owl and groom her back. During the many hours I had spent with Owl, she usually had traveled by herself and experienced limited chances to groom reciprocally. Now she sat facing me, but not seeing me, with Bess hidden behind her. Blondie and Butch were out of sight on the far side of the crown. I was tired of hiding and I stepped into Owl's view.

Surprised at my sudden appearance, she hooed, but immediately her demeanor changed, probably due to recognizing me, and she pant-grunted in the same way she had done when greeting R.P. before they had mated. This so surprised me that I turned around to see if a male chimp, whom I had not seen, was somewhere behind me. I scanned all possible trails, but there was no other ape. I looked at Owl again. She was looking directly at me, and she was *greeting* me! I had never expected it to happen like this, but I had finally arrived.

Not surprisingly, chimpanzees at Ngogo distributed their grooming favors very nonrandomly. Of course mothers groomed their offspring more than they groomed anyone else. But unlike what has been reported from previous studies, at Ngogo adult females without offspring were significantly more likely to groom other females than to groom males. Conversely, males were significantly more likely to groom one another than to groom females. Even when males and females did groom one another, their sessions lasted an average of only about ten minutes as compared to an average of eighteen minutes for pairs of males and more than fifteen minutes for pairs of females. This sexual discrimination was also obvious in the rate at which a partner reciprocated, which was higher among like-sexed pairs (72 percent for pairs of females and 75 percent for pairs of males) than for pairs of mixed sexes (59 percent). And within these most common types of partnerships, it seemed to me that individuals consistently chose specific friends most often.

But, as might be guessed, the hygienic exchange service hypothesis did not explain everything about grooming. Harold Bauer, a graduate student working at Gombe, found in 1977 that grooming at Gombe was more common between males from *separate* parties during a reunion than between males of the *same* party during such times. This suggests grooming has an additional important function, namely to reinforce social bonds and enhance solidarity among males.

In most cases apes who decided not to reciprocate grooming favors possessed the impunity of higher status as, for instance, in the example of R.P. and Spots mentioned above. But for chimpanzees on the low end of the dominance hierarchy, the option to avoid reciprocating seemed almost nonexistent. The following example of grooming not only provides an excellent lesson of this, but also an insight into the evolution of deliberate dishonesty.

All four Farkle family females were basking in the afterglow of figs from a *Ficus dawei*. Fanny, now nearly a two-year old, dangled by one arm while six-year-old Fern tickled her. More sedate, Farkle sat in a wide crotch of limbs and meticulously groomed her oldest daughter. This session already had lasted more than an hour, but Farkle was exhibiting no sign at all of flagging interest in continuing it. Each time it came her oldest daughter's turn to reciprocate, however, a fundamental difference in attitudes became evident.

When her daughter sat behind Farkle to groom her back, instead of diligently parting the hairs and closely examining her mother's skin so she could remove foreign matter or parasites, she lazily ran one hand along Farkle's back in a semblance of grooming while actually gazing at her capering sisters, a passing bird, and ripe figs remaining in the *dawei*. In fact she was not grooming at all but merely making a pretense of it, probably because her mother expected her to groom and she could not "refuse" her mother with impunity. Because this breach of chimpanzee mores seemed so serious, I named her Felony.

What does all this "nonrandom" social grooming mean sociobiologically? A chimpanzee who grooms another does so at some cost of time and energy to itself because reciprocation is sometimes uncertain in quality and quantity. Hence we should expect each ape to generally chose partners in whose health it also has some small but vested interest and/or partners whom it has learned will reciprocate in kind. My hypothesis from Chapter 1 predicted that if male chimpanzees in Kibale were geared for communal territorial defense, then we could expect to see the greatest degree of behaviors leading to increased solidarity between males. With regard to grooming this is indeed the case. The preference of males for one another as partners could have been due to a greater genetic relatedness between them and/or to whatever effect mutual grooming has on increased solidarity. I suspect both factors were responsible, but at this point it was impossible to tell if one was more important than the other. Evidence from Gombe males collected by Harold Bauer and Michael Simpson indicates clearly that grooming between males did reinforce social bonds and acted as social "cement" which may have promoted reliance on one another during encounters with males from other communities. But what about grooming between females?

Females at Ngogo favored one another as partners perhaps because males tended to be nonreciprocators toward them, or because they traveled less with males due to incompatible activity patterns (Chapters 4 and 14), or maybe because their friends were primarily other females, or perhaps for all three of these reasons. Part of the significance of these patterns of mutual grooming, especially among males, is that the social system of the apes of Ngogo was similar, at least in some respects, to that of the killer apes of the Kasakela Community of Gombe. In fact, everything I had seen of the Ngogo males so far had matched with the social patterns of Gombe males, and with the hypothesis that chimpanzees maintained communal territories through male solidarity. But, while grooming at Ngogo was another piece of the sociobiological puzzle that did fit, it was insufficient evidence on its own to make a broad conclusion. I needed more.

To Be Groomed

But I had not analyzed any of this until long after Gray and Zira had set the record for the longest session of mutual grooming at Ngogo. Immediately after this session ended they sat in the tree and simply stared down the hill, not at anything in particular it seemed, but just for the sake of doing something. After traveling, then eating, napping, and grooming together, not much remained to be done—except starting over again once they became hungry. Gray coughed again, a deep raspy rattle that had been bothering her for weeks and would continue to do so.

Zane, who had been sitting a few yards from the two adults for the past hour, returned to the *exasperata* to pick figs from its depleted crop. Forty minutes later she returned to the *Piptadeniastrum*. Earlier she had joined in a grooming chain with Gray and Zira; even Owl had groomed with this trio yesterday. I anticipated Zane to spur another grooming chain, but though her return aroused the two older females from their torpor, instead of grooming again, soon all three climbed down and abandoned the remaining figs. I wondered what prompted them to go.

I also wondered about Zane's anomalous relationship with the two older females. She was only about seven years old, too young to always be away from her mother. Gray was certainly old enough to be her mother but she did not act the part, being much more attached to Zira and traveling with her four times more often. She also was old enough to be Zira's grandmother, which I suspected due to the occasional closeness between the two, impressive because Gray snubbed other juvenile females. Zira was only about fifteen years old, if that, far too young to have a seven-year-old daughter, and she acted the part even less than Gray. She did not even act close enough to be Zane's sister. I finally concluded that Zane was an orphan whom Gray had befriended but not quite adopted, but because such a relationship would be unusual among chimpanzees, I continued to suspect some familial tie. I continued to watch for clues.

Because I had been riveted on them for six hours, it was a relief not to be scrutinizing their every move. But, even so, I found myself wishing they had stayed longer.

The sky had become overcast and the air still and heavy. I could almost feel an electric tenseness in it. The forest was now abnormally quiet; I began to feel uneasy for no definable reason.

I was chewing on a handful of groundnuts and suddenly sensed their presence. Maybe my nose told me, maybe my ears: I do not know how I knew they were back, but I knew. With a quickening heart I scanned through the saplings and vines silently battling for more sunlight and saw a slow movement about thirty yards upslope. I lifted my field glasses for a closer look at an arm apparently reaching up from behind the foliage. Although I had not seen it clearly, my mind's eye pictured a chimpanzee, but the arm snaked upward with unnatural suppleness. Next to it was a gigantic ear the same shape as the continent on which we were standing.

I turned toward the tree I had found earlier and was about to pass the camera balanced on my pack on the ground. Delbert True, chairman of the anthroplogy department at U.C., Davis, had reluctantly authorized me to take it. "If you fall into a swamp," he had said in a tone revealing little hope that I would not, "hold it up in the air."

I looped the strap around my neck so the camera hung down my back then ran to my tree and clambered twenty feet up. Before I made it that far, though, fifteen elephants crashed and smashed a path twenty-five yards south of me, while simultaneously four others tore through the understory about fifteen yards north. Had I spooked them? I unslung the camera and aimed it at the rushing red-mud-coated bodies, but due to the dim light, the film developed later as a red streak. As before, both lines of elephants slid to a last second halt at the brink of the chasm, then diverted at right angles. Less than thirty seconds later five or six more rumbled after the larger group.

This seemed too much like the Africa of Edgar Rice Burroughs. I was perched in a fragile tree while elephant after elephant stampeded past me. A moment later a single elephant trumpeted piercingly from across the ravine. Were poachers spooking them? They had to be.

I was glad to be in that tree, so glad that I did not climb down for an hour. It was easy to see why chimpanzees favored trees as resting places: security and a good view. More mosquitoes though. After a half hour's absence, Gray, Zira, and Zane returned to the *exasperata* along a path under my tree. Had they avoided the second elephant stampede by coincidence? Zira looked up and saw me. I had entered a new dimension by leaving the ground. Reverse evolution. Apparently this was too much for her; she bolted. Zane and Gray glanced up and then followed. Perhaps they did not even recognize me up there, but it was too late. They were gone. Soon a solitary blue monkey passed nearby and entered the fig tree, but after two minutes he spotted me and rushed out. Their message was clear: I belonged on the ground.

I returned to my usual spot for the final couple of hours of this vigil. Elephants were stomping along worn furrows in my brain. Was I pushing my luck by staying?

I was reluctant to pass up any opportunity to observe the apes. This *exasperata* was the first major food tree I had found in over a month. Currently the majority of primates, including the apes, were harvesting the tasty fruits of small *Uvariopsis* trees, but finding the chimps in the scattered groves was difficult. According to my phenological studies, the amount of fruit at Ngogo was now in its lowest ebb during my entire research; the dry season had dispersed most of the apes to parts unknown. Before finding this *exasperata*, I had not even seen an adult male for a month. This fig tree had attracted seventeen chimps back from wherever it was they had gone. Though my desire was strong to remain for these last couple of hours, the urge to avoid yet another stampede of elephants was even stronger. I was spooked.

Camp was only a mile away, but never had the forest seemed so packed with surprises. I walked as if on eggs. And I wished for X-ray eyes as I studied the foliage along the route. Each bird fluttering from the understory and every squirrel scurrying up a limb set my heart to racing. That mile seemed to take forever.

"*Tembo mingi sana* (very many elephant)," Benedicto and Rhawire pointed out to explain why they had done no cutting all day. They grinned sheepishly and scuffed their bare feet. Remembering my own experience that day, I found it hard to chastise them.

The now barren *exasperata* forced me to extend my searches beyond the trail-grid system. Because to the east and north the forest was less extensive and mixed with forests of less diversity plus colonizing scrub and elephant grassland, I explored to the south and west where the Ngogo chimpanzees had repeatedly traveled in the past. Most of this territory was new to me and I tried to memorize its features as I searched for the apes, their nests, and trees in fruit. These regions looked no better than what I had left behind, but undoubtedly the apes knew the potential resources here better than I did. In fact, this dry season was about to reveal to me a new facet of chimpanzee sociality. My routes cut through dense thickets of spiny *Acanthus* riddled with elephant paths, which in turn showed slashmarks from *pangas* wielded by poachers. No wonder the elephants were jumpy.

One afternoon I returned to camp to hand crank Old Rover into life to recharge its dying battery, an obeisance it demanded more and more frequently. As Otim approached to help, I looked beyond him toward the big hill and saw a spiral of smoke rising into the hazy sky from the far side of the ridge. As we stared, poachers filed to the ridge line one by one until we thought there would be no end to them. I counted twenty-eight hunters bristling with spears and carrying at least a dozen nets. As if contemptous, they stood sky-lined by the smoke and surveyed our camp. Mukasa joined Otim to hurry along a forest trail to intercept them. But because they had only four bullets between them, I was sure they would return empty-handed.

Crackling flames crested the hill an hour later. The wind was northerly, away from camp, so I guessed the conflagration at best had only a fifty-fifty chance of reaching us. My mud hut would not burn, but the thatched roofs of the other three would ignite like tinder boxes. Unfortunately the sward extended only ten feet beyond those huts. Benedicto began hauling the meager possessions of the Ngogo crew into the open so they would not be lost when a burning roof collapsed. Then he started hacking at the elephant grass to expand the fire break. Two hours later Otim and Mukasa returned empty-handed.

Flames leaped twenty-five feet into the sky and crept down the hill toward us. After nightfall the crackling blaze illuminated a wavering sheet of smoke looming over camp like a ghostly specter. I sat on the sward and watched its progress. Unless the wind veered toward us, the conflagration would not overrun the compound for many hours. Hopefully, the sward was too short to burn. I went inside, laid back on my bed and listened to the popping flames. I watched for hours through a window. By the time I fell asleep my mind was dominated by fire, and the flames crept closer in my dreams.

During the night the fire died about 400 yards from camp. I entered the forest to search for other chimps, found none, and returned nine hours later. During my absence and that of Otim and Mukasa who were on patrol, poachers had returned to the big hill to kindle a new blaze further south, which already was much closer to camp. I found it difficult to interpret this as being motivated by anything other than malevolence directed at us, the tiny but inconvenient shred of authority at Ngogo.

Night fell and flames crackled. I hit the sack and wondered if this time the tongues of fire would lick far enough to give those thatched roofs the kiss of death. Eventually I slept and my dreams mirrored the hillside. Meanwhile the fire died again in early morning. Within a month the elephant grass regrew seven feet high on the big hill. Poachers did not put it to the torch again for six months, but that time the poachers would not escape (Chapter 11).

This dry season was revealing another perspective of chimpanzee social life. During these months of few figs and little fruit in general, the apes I found were solitary much more frequently or traveled with just a few companions. Pant-hooting was now very rare, as if fruit itself was the only reason for these vocalizations. The dry season, and its reduction in fruit, was the weak link in the chain of chimpanzee social life. Competition for food seemed to make socializing so expensive that the apes were on a temporary austerity program—social grooming was now a greater luxury than usual. But while observations were difficult, the apes were revealing new behaviors that broadened my understanding of them.

Finally, three weeks after the figs in the *exasperata* had run out, the beginning of August ripened the fruit of a large *Cordia* tree I had been monitoring in the south. I staked it out in anticipation of my first customers. The rustle of dry leaves alerted me to a thin, unchimpanzee-like hand reaching up through the foliage. Its owner, an adult male baboon, followed. Three seconds behind him came S.B., whom I had not seen in months. They moved to opposite sides of the tree, following chimpanzee procedure, and fed on the sticky, datelike fruits. Apparently they had been traveling together—a very odd couple.

Anubis, or olive baboons, were fairly common at Ngogo; at least two troops foraged within the trail-grid system and accounted for 30 percent of the local primate biomass. But despite being common, I often spent entire weeks in the forest without seeing them because of the obscurity of the dense vegetation, which required well over sixty miles of walking to actually see the entire area within the trail-grid system. (At any given place, usually less than one thousandth of the two square miles was visible.) As with chimpanzees, these terrestrial monkeys' prior experience with hunters had probably instilled in them what was now a culturally promulgated fear of humans that made them difficult to observe and, as Simon concluded, impossible to study. But also like the chimpanzees, each successive encounter now left them slightly less prone to flee. I observed them at every opportunity.

For primatologists of the 1960s baboons provided the surrogate for societies of protohumans roaming the predator-haunted, primeval savanna. In open terrain troops were easy to follow in a vehicle, and, once they became blasé about that, investigators started standing outside it, then abandoning it altogether to follow the monkeys on foot. Dozens of projects have elucidated the fascinating details of the lives of Africa's most successful nonhuman primates.

Like most social primates, baboon societies are matriarchial. Craig Packer reported in 1975 that Gombe males matured and emigrated; Glenn Hausfater and Jeanne and Stewart Altmann reported that females at Amboseli mature and remain in their troop to inherit much of their mothers' status, and to give birth to females who do the same. Immigrating males sometimes follow in the footsteps of older brothers, and often gain acceptance into a troop by forming what Shirley Strum described in 1975 as a "special relationship" with an adult female, which might or might not include mating privileges. Packer (1977) found that, once in, males usually formed alliances with one or more other males for mutual support in defense or offense against other more dominant males, or other alliances to gain higher status and more opportunities to mate. Robert Harding, Sherwood Washburn and Irven DeVore, and others reported that males also tend to act as vanguards during travel. Strum (1981) observed the development of a hunting tradition among Gilgil males, who, by detouring into places where female antelope sequestered their young, became the primary hunters. But the females of the matriachy seem to be the troop's repository of knowledge and the keepers of tradition. This portrait was gleamed from several thousand hours of observation in open habitats. Thelma Rowell made a pioneer effort to observe anubis baboons in a narrow gallery forest along the Ishasha River in Queen Elizabeth Park. Despite the virtually limitless expanse of savanna, the baboons spent 60 percent of their daylight hours among the trees.

Rowell's observations suggest that either baboons are a forest species adaptable enough to make good in the savanna, or simply a terrestrial species opportunistic enough to make good in any habitat where their basic needs are met. Their adaptability suggests the latter. Almost certainly Rowell's baboons were safer from lions in the gallery forest than on the savanna (Chapter 13). Mysteriously in Kibale Forest, no troops frequented the study area at Kanyawara, but the baboons of Ngogo spent most of their time in the forest except immediately after the grasslands burned, and they were the healthiest baboons I saw in Africa. No one had any solid information on them, or on any other true forest-living baboons, and I often wondered if their social structure differed from that of their cousins on the primatologist-haunted savanna.

The male baboon with S.B. was probably shopping for a new troop that would offer him the opportunity to attain higher status. His temporary partnership with S.B. was puzzling. Did they mutually groom? As a whole bevy of questions suddenly occurred to me, I saw the baboon carelessly drop several sticky seeds on his own coat where they adhered like food on the face of an oblivous drunk. If they did groom mutually, the monkey was getting the better end of the deal; S.B. was, like all chimpanzees, fastidious. Unfortunately the baboon soon spotted me, rushed out of the *Cordia*, and cut off my flow of clues. A few seconds later he barked out a warning but S.B. paid no apparent heed.

During the ensuing week at this *Cordia* I saw only eight chimps, including Gray, Zira, Newman, Eskimo, and Shemp, but at the end of it I received a bonus in my education. At the creek flowing fifty yards south of this *Cordia*, I found R.P. sitting on the ground next to a rotting log. He half leaned on his left side toward the log and poked at it. He sat up, lifted his right index finger, examined it, then put the tip of it in his mouth to eat something on it. I had to suppress an impulse to go sit next to him. Instead I merely spent 20 minutes edging closer. He gazed toward me and gave occasional deep "hums," but each time returned his attention to the log. He briefly groomed his left forearm, then again studied the log. After several minutes he stood bipedally, swayed once to each side, ran bipedally past me toward the *Cordia*, and thumped a tree bole in passing, his display being aimed obviously at me. I examined the log. It was about eighteen inches in diameter and a few yards long. About 2 yards of it had been torn up; four sections had finger holes freshly dug into the soggy wood. With my hunting knife I excavated a few termites R.P. had missed.

This was my first observation of chimpanzees eating strictly animal foods. Because, unlike most fruits, figs normally contain wasp larvae and important additional protein and B vitamins, for much of my time observing the apes they did not need to seek strictly animal foods. But now, when few or no figs were to be had, the available fruit contained too little of both, and the chimps were forced to seek alternate sources of protein.

Termites in Kibale did not build mounds, and underground bees' nests were very rare. Consequently, the apes here had little incentive to contruct and use honey daubing sticks as observed by Fred Merfield in the 1950s or termite "fishing" sticks reported by Jane Goodall in the 1960s, or the ant fishing sticks reported by Toshisada Nishida a little later in 1973. Termites here lived in dead wood and had to be dug out. At Gombe chimpanzees also hunted mammals when opportunities arose and exhibited varying degrees of success. William McGrew reported in 1979 that Gombe males were usually the hunters while females were normally the expert termite "fisher" chimps. Potentially females are nearly as good on the hunt, but during much of their adult lives they must transport an infant whose safety is incompatible with the arboreal scrambling demanded by most hunts. A long time passed before I was to see it, but the apes of Ngogo also ate monkeys (Chapter 9).

After quitting his termite mining, R.P. joined Newman and Shemp in the *Cordia*. As if firm adherents to a manual of chimpanzee table manners, none of the males interacted directly. They foraged until twilight when R.P., then Newman and Shemp, descended to the obscurity of the forest floor.

When the beginning of September finally brought the rains the chimpanzees' austerity program ended. A large *Ficus natalensis* near camp produced a huge crop and became a local hot spot. By this time entire weeks had passed during which I had searched the forest daily but found no chimpanzees. It was a relief to finally connect again, but it was at this tree where I nearly lost permanently one of my best allies as the price of a history lesson.

Owl was alternately foraging in the *natalensis* and chasing large black and white casqued hornbills out of it. Zira and Gray walked past me only a dozen yards away, gave me a brief glance (I was not in a tree), then carefully climbed up to join her. For the next three hours the females fed, groomed one another with obvious relish, built day nests and napped, then fed again within the neighborhood of the fig tree. This sedate life-style of females clashed with the more purposeful, almost harried routine of males, and it was fine with me; I needed a rest. By afternoon the sky clouded over, flashes of lighting cut jaggedly from the heavens, and thunder pounded our ears. None of the females seemed to pay any attention to the flashes and roars.

Suddenly a fork of lightning materialized above the *natalensis* as an instantaneous inverted bare tree of electricity hovering but not quite touching. Thunder was simultaneous and deafening. Gray flinched and turned quickly to scrutinize me intently for several seconds. Then she abruptly rushed out of the fig tree and climbed 100 yards away to actually hide from me. Gray clearly wanted to be out of my sight but did not want to abandon Zira. Neither young female paid any attention to the lightning, thunder, me, or even Gray's behavior. They continued to pluck figs for another hour.

From the instant this bolt of lightning struck and for months afterward Gray fled from me on sight. That she had convicted me mistakenly on circumstantial evidence bothered me, but more important, I feared that I had lost her. After the lightning she responded to me as if I were the devil, a total reversal of our previous relationship. Professionally her new distrust hurt my observations, but on a deeper level I felt jilted unfairly, a victim of mistaken identity. Soon, though, I realized she was not being unfair; as a human, I was guilty.

Until fifteen years before this episode, Bajonko poachers had hunted in parts of this forest for primates and all other game. Yongili told me he had heard that some of them had used firearms. If Gray, who certainly had been an adult then, had witnessed the killing or wounding of other chimpanzees by the guns of the Bakonjo, or even had been a victim herself, she might have associated me and the thunder with those memories. Neither Owl nor Zira had been born during this era of hunting. Does this explain why they paid no attention during Gray's obvious fright? Of course it is impossible to know for sure, but I became convinced Gray thought I was trying to kill her. My main worry was how to redeem myself.

In the darkened interior of my mud hut I read a letter from John Yost, vice president of Sobek Expeditions of Angels Camp, California, "You're on the first trip if you want it. This requires showing up in Ethiland [sic] about Sept. 20. Of course there is no guarantee that the trip will go; politics there are almost as bad as Uganda."

Before arriving in Uganda I had paid my bills in graduate school by working as a professional white water guide on wild rivers. Africa had offered no such opportunities, nor had I yet received a single grant to help support my research, but now Sobek was offering me a position on the new run they had pioneered in Ethiopia. The Omo River, the "African Queen," had been floated only a few times but its reputation already was drawing commercial clients from the United States. Inconveniently, Ethiopia itself was tumultuous. Most of the American Embassy, plus a few European embassies, already had fled the new socialist, pro-Russian regime in Addis Ababa. Radio Ethiopia had been broadcasting scathing anti-American-capitalist-imperialist diatribes leaning heavily on clichés like, "oppressed landless masses rescued by revolutionary cadres of the new socialist movement." Ethiopia was also bogged down in a hot war with Somalia in the Ogaden Desert and was apparently losing. I would be there in a few days.

After having spent fifty-seven days on another thirty-day leave, Yongili returned to Ngogo with his brother who needed work. I organized a schedule to ensure that all those *pangas* would rescue the disappearing trail-grid system during my absence. I drove Old Rover for a huge food run to Fort Portal for the Ngogo crew, then made a list of things I could bring them from Kenya on my return. Then I put my gear into Old Rover and hoped it would make it to Fort Portal again. Because Struhsaker, now in the U.S.A., would return to Uganda with a new vehicle from Kenya,

this was Old Rover's final run from Ngogo. As a parting statement the entire exhaust system jettisoned onto the road.

After a hair-raising ride by taxi to Kasese Airstrip and two hours
of turbulence in a Piper, I found myself waiting out the night at
Entebbe International for a dawn flight to Nairobi en route to
Addis Ababa. I could not afford a hotel room, so I stayed in the
departure lounge.

I was gun-shy in the political center of Uganda. Four months
earlier, after having dropped Simon off, I had been placed under
house arrest at Makerere University by a fat, bejewelled, high-
echelon agent of the State Research Bureau. Because he was
illiterate, I was detained twenty-four hours while an underling
read my passport and identity cards to him. A replay of that was
the last thing I wanted now.

I pulled off my boots, found six chigger egg sacks embedded in
my feet, and dug them out with my knife. Then I leaned against
my North Face pack and tried to sleep as guards armed with
machine guns paced the room all night. Soon a jet would take me
out of the frying pan into the fire.

8

River into Time

My boat drifted as if attached to a moving milk chocolate mirror framed by cliffs of basalt dropping half a mile into the African crust. We were among the first hundred people to ride the surging swell of the Omo River. Many of the tributary canyons we passed had never known the tread of a boot; some probably had never known the tread of any human foot. Until well into this century, hundreds of miles of the Omo had been represented on maps as terra incognita up to where the river writhed python-like into the region known as the Plain of Death. The survey work was considered suicidal.

In the days since we had put our five boats on the river at Gibe Bridge, the rhythm of its waters had become second nature. Constrained within a gorge it had carved over millions of years, the Omo churned, twisted, boiled, roared, cut, then grew calm— as if confident to overcome anything in its path. Its canyon was home to many of the wild creatures that are Africa's vanishing treasures, including some old friends from Kibale Forest: waterbuck, bushbuck, anubis baboons, and black and white colobus.

We drifted as still as statues. A one-male harem of colobus appeared among the emerald lianas clinging to the black basalt.

Because we had not moved, and possibly were not recognizable as humans, the monkeys merely returned our gaze with wary stares instead of fleeing. Two black juveniles leapt from limb to limb in a game of tag, their long white fringes of guard hairs flowing like barbaric ornamentation donned for some primitive ritual. Their elders watched them stolidly. Unlike most primates, black and white colobus spend most of their daylight hours doing nothing. Their bulging sacculated stomachs allow them to feed on common leaves, which are easily found and harvested, so they are satiated by short bouts of feeding. They spend the rest of their time sitting and digesting like cows without cuds. The torpidity of these monkeys seemed to give license to our own. This moment required nothing more than being alive. I was almost sorry to hear the increasing roar of the upcoming rapid; this moment with the colobus would end too soon.

WHOOSH!

A bull hippo surfaced only four feet from the boat! The silty waters of the Omo were too turbid for the hippo to have seen us from below, and no one on the raft had made a sound that might have warned him of our presence. Now we were inside his margin of safety.

A large hippo can rip an inflatable boat to shreds. Shredded it would sink. Being in the water with an aggressive bull hippo is bad enough, but the Omo is also home to some of the most notorious man-eating crocodiles in the world. The thought of swimming it just above a rapid with a bull hippo and possible crocodiles waiting below was anything but pleasant. Would he attack or flee? We were too close for me to escape by rowing.

The great river horse blinked as if unbelieving, turned his massive head and keen scythelike tusks away from the raft, arched his back, and rolled like a whale to vanish beneath the water.

My relief was short-lived. Now I had to position the boat for a good entry into the rapid. "Whew!" I whispered half to myself and half to my passengers, Barney, Ward, and Margie, "He was *close!*" Somewhere in the murky depths beneath us that hippo was caught in the current feeding the roaring rapid ahead.

A silky smooth, V-shaped tongue of water narrowed to a point ahead of us. I scanned the rapid again for hidden dangers. The bull popped up ahead and to the left, looked around, saw us, then submerged. He was about to run a respectable rapid submarine style. I started worrying about him.

The eighteen-foot river boat surged up the first swell of a series of haystack waves leading downstream like a roller coaster. I tried to imagine what was happening to the hippo in the maelstrom below us. Below the surface of a large rapid lies a dynamic world of irresistibly powerful currents; a submerged human is helpless and disoriented. Hippos probably do not feel at home in such turbulence either. I felt a pang of guilt over having frightened this Pleistocene river horse into being pounded by the relentless conflict of water and rock below.

He appeared again only ten yards ahead. His great-jawed face peered from the white water for a second, then disappeared. Normal swimming technique for hippos involves bouncing off the bed of the river in a kind of zero-gravity, slow-motion moon walking: a graceful performance when viewed through clear water. Perhaps this beast *was* at home in a rapid and not in real danger. He appeared a last time then submerged. I suspect he maneuvered his bulk into the eddy at the foot of the rapid as we were carried downriver.

We were members of the tenth Sobek expedition down the Omo. Previously, two ill-equipped attempts by West Germans had been aborted, and a National Geographic–sponsored expedition had raised official eyebrows in Addis Ababa by departing Ethiopia via Lake Turkana, the Kenyan lake fed by the Omo. Though our expedition of twenty-one members was not the first, now the flow level was near a record high, run only once before. And because so little of the canyon had been trod, despite previous expeditions, it seemed like a voyage of discovery.

Ethiopia sprawls across a mountainous plateau isolated on the Horn of Africa. Surrounded by searing deserts, the peoples of this plateau have had little contact with the civilizations beyond it for thousands of years—until this century. The miracle of aviation has infused the endemic cultural diversity of Ethiopia, rivalled only by its topographic diversity, with new, predominantly Western ideas. Despite the delivery of Western culture by plane, rural Ethiopia appears little changed even under the new socialist regime. Much of the tortuous terrain has yet to feel the gnawing of road graders or other heavy construction machinery, and by Western standards it is still isolated. The canyon of the Omo is one of the most back-of-beyond regions of this back-of-beyond nation. We were beyond the eyes of authority, and our mistakes beyond the assistance of the outside world. Almost as bad as Kibale, a serious injury here could mean death.

But among our fourteen clients were three medical doctors, two dentists, and a pharmacist. After spending nearly a year in Uganda without seeing a doctor, this expedition seemed like a medical convention.

Only ten days ago the paying members of our expedition had strolled from their hotel under the bright lights of Addis Ababa to dine on traditional food in a fine restaurant. This had been the first opportunity for all of us to meet; we Sobek boatmen, Jim Slade, John Yost, Bart Henderson, Dave Hinshaw, Bob Whitney, and I, had been running errands for dozens of last minute supplies and had been scrambling to repair equipment stored from the previous year, so this was our first evening off. We celebrated by consuming impossible quantities of wide, floppy sourdough *injera* (flatbread) flooded with a variety of spicy stews called *wat* and by drinking enough *tej* (honey mead) to break the local record. Our Ethiopian version of a Roman bacchanal was memorable, and we paid for it again the next morning. Alone it may have been worth the rigors of travel to Addis Ababa.

Now, far from the neon lights of Addis, we were in another world; the day-to-day life on the Omo was the only reality. At 6:30 the sun dispelled the Stygian gloom of the cloudy night. The neighborhood bull hippo who had snorted and bellowed his territorial challenges through the darkness became quiet as we emerged from our rain-soaked tents to place half a lit candle in the fire pit and slowly feed in wet tinder until it burned on its own. Then we brewed Ethiopian coffee, the best in the world. Ethnobotanical evidence, in fact, points to the drinking of coffee originating near this region of Ethiopia. After eggs, pancakes, crepes, or granola, we struck tents, loaded the rafts, and rowed to midstream on the swell.

Family groups of hippos had already collapsed into heaps of rotund, grayish pink bodies with bewhiskered snouts propped above water on one another's backs. Their collective bellies bulged with hundreds of pounds of grass cropped during the night. Some slumbered submerged, some on shore, and some with portions of their bodies in both places. They seemed oblivious and secure. Often they did not notice us drifting by unless someone made a noise or yelled, "Hey, Hippo!" just for the hell of it. Their sudden awareness of us precipitated an immediate, half awake plodding to the water, where they vanished as if they had never existed. Eventually, after we had drifted further on the spate, pairs of nostrils and eyes surfaced in the eddy upstream and watched us disappear.

Mottled green crocodiles lunged into the river as we came into view. These most successful remnants of the age of dinosaurs were more wary than the hippos and usually took to the water to escape rather than attack. Small groups of elegant waterbuck stepped along the cobbles at the water's edge. Upon seeing us they melted into the riverine vegetation beyond. Pairs of white-headed African fish eagles gazed down on our flotilla from perches in twisted strangler figs. Mist diffused to reveal craggy cliffs of jet, festooned with lianas. No human eyes other than our own were witness to these Pleistocene mornings. Our rafts and the Omo were a time machine.

Before lunch (peanut butter and jelly on camp-baked bread that inevitably failed to rise, British cookies with staleness baked right into them, plus fruit), we usually stopped at a tributary creek for fresh water, to explore, and to bathe. Some of these were small rivers, brick red or brown with sediment carried by rains from mountains and the plateau farms. Others were crystal clear. A few were so sluggish they harbored the snail that plays host to the larvae of bilharzia, a schistosomal worm that burrows through one's skin, into the circulatory system, then tunnels out through the lungs, crawls up the trachea to the esophagus, and finally slithers down into the stomach. After this incredible journey the flukes burrow into the intestinal lining, where they produce hundreds of eggs per day. Infestation produces night sweats, fever, and intestinal pain and, if the infection is heavy and goes untreated, can be fatal. Slade, our trip leader, and I did not discriminate keenly enough at one of these creeks. He dived in while I filled a water jug; almost immediately he felt the telltale swimmer's itch. Months later he had bilharzia, but that time the flukes missed me.

From the river we watched waterfalls cascade up to 1,000 feet through space to diffuse into tatters of mist. But not all side creeks were awesome. Some were sybaritic delights created by

Mother Nature during sensuous moments. At one, Fuller's Flume, a warm clear creek gurgled from one sculpted bedrock basin to another in a terraced series of natural baths equipped with water massages, fringed by golden grass and yellow-blossomed acacias. Tearing ourselves away was difficult.

Hundreds of miles of the Omo River canyon is an unspoiled remnant of wild Africa, a paradise of sorts. But it is a paradise guarded by a demon. Or rather by thousands of them—small flies that cross their wing tips when they alight. Tsetse flies look innocuous, but bite like a hypodermic in the hands of an army medic. Tsetses on the Omo were out for blood and impossible to discourage. A direct hit with an open palm that would squash insects of a similar ilk was merely an inconvenience to a tsetse. Time after time I smashed one of these little demons just before it turned me into a fast-food stop, then watched in grim satisfaction as its corpse toppled into the bilge of my boat. Soon an amazing reincarnation occurred: a few twitches upside down in the water, a buzz or two, a violent jerk to right itself, and suddenly the damned thing flew out of the water, alighted on my ankle, and tried to bite me again. They were unbelievably sturdy and persistent. And despite wearing long pants and long-sleeved shirts, I collected hundreds of bites.

Tsetses have been libelled the "bane of Africa" because the various species of tsetse (*Glossina*) transmit five forms of sleeping sickness (*trypanosomiasis*), two of which are fatal to humans. Though today only about 20,000 new human cases are reported each year, Uganda's population was cut in half by trypanosomes early in this century. Three other forms of the disease kill domestic livestock. Biologists John Donelson and Mervyn Turner reported that in more than ten million square kilometers of Africa (an area as large as the United States) tsetses and trypanosomes prohibit keeping livestock, except for very low-yield strains of Zebu cattle, and sometimes not even these. The amazing success of trypanosomes is attributable to their unusual ability to continue changing the molecular structure of their coats, thus foiling their host's antibody responses much as does the AIDS virus. Except for a few relict tribes of hunters, such as the Hadza of Tanzania, the Bambuti of the Ituri Forest west of the Mountains of the Moon, and the !Kung and San bushmen of the Kalahari Desert, all rural Africans keep livestock. By carrying their invisible but fatal protozoan message, tsetses safeguard much of the remaining wilderness left in Africa by discouraging human settlement.

But these baneful guardians do not reign uncontested. Tsetse elimination programs have been conducted over many regions of the continent. The simplest of these consisted of paying people to act as bait, scour the bush, and catch and kill the flies. Because adult females give birth to only one nearly developed larva at intervals of several months, an infested region never has many flies, although their persistence makes it seem otherwise. Partial removal of tsetses can appreciably reduce a local tsetse population, and their low birth rate prevents rapid regrowth.

One scorched earth program applied extensively in Rhodesia and western Uganda involved shooting all wild game to remove the immune reservoir for trypanosomes and the food supply for the flies. Attempts like these have failed because both trypanosomes and tsetses survive in and on small mammals nearly impossible to extirpate completely. Scandalously, tens of thousands of antelope have been left to rot on the ground in these campaigns in protein-starved Africa.

A third approach relies on interfering with the female flies' reproductive needs, specifically shady bush in which to deliver their offspring. Clearing of dense bush in Nigeria successfully reduced fly numbers. But, of course, clearing woody plants from a wildlife habitat alters it significantly and eliminates many other species of wildlife.

The most recent and modern approach, used extensively in Zimbabwe, now employs traps set in the bush and baited with an attractant chemical most tsetses find irresistible. Apparently these traps are virtually clearing out regions infested by the fly, and nothing else is affected directly. Trapping, however, is a preliminary to human invasion of the wilderness. But along the Omo River none of this has happened and the tsetse *does* reign uncontested and oblivious to the fact that it alone protects this wilderness from human settlement.

Despite the fly and the sleeping sickness, some Ethiopians hiked down side canyons to the Omo to water their cattle, to cross the river to reach a market on the north side of the canyon, or to bathe in hot springs near the river. But the herders never kept their cattle permanently near the river despite the availability of forage.

Because of the danger of crocodile attacks, we never swam in the Omo. There were times when the heat was so oppressive that I dropped the oars, scanned again for crocs (a constant habit), then plunged into the river for an in-out dip, but I kept these to a minimum.

One day early in the trip, as we stopped for lunch a Walyta (the predominant people inhabiting the plateaus here) leading a donkey burdened with two large sacks appeared on the far side of the river. He stopped at the water's edge, unloaded both sacks and a large transistor radio, then inflated a pair of goat skins whose leg holes, etc., had been tied to make them airtight. He tied the inflated skins on each side of his donkey and led it to the river. The two swam across, the man pulling the beast, and reached the opposite bank considerably downstream. The aquatic Ethiopian hiked back upstream with one skin and swam across the river to retrieve his load. He balanced one huge sack on his head, straddled the goat skin in the water, and managed to paddle swim across safely to his donkey. He returned a second time to pick up the remaining sack and paddled back again without incident.

Where were those man-eating crocodiles? So far we had seen crocs daily, several ten feet long, but no monsters. But even a ten-foot crocodile could attack a man. I had been keeping a count of all the large animals we saw while floating downriver. Hippos on these upper 240 miles of the river were quite numerous, but I counted only one-twentieth as many crocodiles and they were relatively small. Perhaps the more turbulent water of the upper river was a poor area for them to hunt. Along the final 90 miles of the Omo, upstream of its confluence with the Mui River, the hippos had been decimated by local hunters but the crocodiles were bigger, bolder, and far more numerous. Tribesmen there were emphatic about man-eating crocodiles and never went swimming. Later I learned that even the Walyta who swim across this upper stretch are attacked occasionally, but, having no bridges or boats, they consider the swim unavoidable and the attacks a calculated risk—Omo roulette.

Nile crocodiles are well-rounded predators that have existed essentially unchanged since the extinction of the dinosaurs sixty-five million years ago. Crocodile biologists Anthony Pooley and Carl Gans reported that males may exceed seventeen feet in length and weight close to a ton. Nile crocs, like these on the Omo (which a few million years ago was a tributary of the White Nile, before geological uplifting cut it off in the region of Lake Turkana), can pull down a buffalo of their own weight. Far from being the sluggish sit-and-wait scavengers that most people imagine loiter on river banks waiting for dead or disabled animals to float by, Pooley and Gans found that crocs were active, efficient predators who sometimes cooperated on hunts and in dismembering large prey. We passed hundreds of them on the Omo. I would have attempted to swim across the river only if the alternative on my side was certain death.

We took precautions: especially on the lower river, none of us went near the water after dark. A flashlight beam sometimes revealed an eerie scene: a half dozen pairs of glowing red eyes staring toward us from a crescent formation around our moored boats. The distance between the eyes of a pair gave a good idea of the saurian's size. Our caution served us well; Sobek (named in honor of the Egyptian crocodile god as an obeisance to the guardians of the river) had experienced good luck in regard to the real thing. The only croc attack before this trip had been aimed at a boat; no injuries.

Within the widening canyon of the Omo were people so far removed from Western civilization that its total obliteration would not have caused a ripple in their pond. Every few days since the start of our journey we had contacted splinter groups from different tribes. Communication was rudimentary and would have been nearly impossible without the aid of Seleshie Haile, an American-trained biologist assigned by the Ethiopian Wildlife Conservation Department to our expedition to census crocs and hippos (our independent counts for hippos during 330 miles were 1,160 for me and 1,186 for Seleshie). He spoke American English, French, Italian, and, of course, Amharic, and he became an essential member of our team.

More than 200 miles into our journey, in the midst of a back-straining marathon to reach a place called Hippo Falls, we stopped to hike to a hot spring about ten minutes from the river. Slade led us along a sinuous hippo path ascending a side canyon. Two old women, an infant, a boy about ten years old, a very old man, plus a few other men of the Welamo tribe had arrived from the rim ahead of us and were stripped to bathe under a man-made spout gushing a stream of tolerably hot water into a bathtub-sized pool. Five more men arrived from upslope a minute behind us. Two were afflicted with ghastly running sores symtomatic of yaws, while a third's scabby hands were frozen into permanent useless claws. Slade and I shook hands with everyone and muttered, "*Salaam*" and "Hello." The newcomers politely joined the circuit to shake their right hands with ours, while using their left hands to cover their genitals.

Not one healthy specimen in the bunch, I mused ruefully, but I suppose that is why they were here. But it did not speak well for the efficacy of bathing in hot pools. Unfortunately, our trip doctor, Charles, had not brought medicine from the boats. Penicillin would have cured those nasty cases of yaws. Rick and Dave stripped to bathe under the spout. I watched my step; human excrement was scattered within a few feet of the pool. Apparently these Welamo had no notion of sanitation or its connection with health. Excrement, ignorance, and sickness in paradise. My right hand started to tingle. I could almost feel virulent microbes trying to hammer their way past my pale skin so they could go crazy inside my vulnerable foreigner's body. I resisted an almost over-powering urge to scrub it until later, but I was careful in the meantime not to touch myself with my pariah hand.

Soon we left the ailing Welamo to again row downstream. I was sorry we could not do more for them. I rowed ahead of the other boats and rounded a bend to see five men on shore standing over a hippo whose flank was dissected by a bloodly red line. Two of the hunters had old bolt action rifles and ran along the shore to shout at us. The hunters sprinting past the crimson river horse seemed a vision from some primal age now extinct. They wanted me to row ashore, but we were running out of time and I was not in the mood to shake any more hands that day. I continued downstream and left the pleasantries to Seleshie in the boat behind mine. He told me later that the men were militia, with government permission to hunt for subsistence; they had wanted to know our business in this remote region of their country.

By late afternoon we reached a small beach at a clear tributary, then hiked less than a mile from the river to where Hippo Falls plunged eighty feet into a huge pool in solid bedrock, a true fragment of the original paradise which we enjoyed to the hilt. We camped on that beach that night, the next day, and the next night.

Two days downstream from Hippo Falls, during the final three days of our seventeen on the river, we entered the Iron Age. The canyon had widened dramatically. Gone were the lofty, mist-shrouded crags, precipitous slopes, and roaring rapids. Now the Omo meandered lazily southward in a widening plain dotted with oasislike clusters of palms. Beyond the narrow belt of riverine vegetation was scrubland, woodland savanna, then true savanna. The highlands were now many miles north. South of us the Omo watered an ever drier land, one which had yielded some of the oldest and most important fossils of ancient and protohumans found to date.

We were drifting on a wide river denuded of hippos, but well stocked with crocodiles, when we spotted the first members of the first of three Nilotic tribes which lay ahead. Where the steep bank levelled off ten feet above the river, an old man and his two grown sons worked to clear a small field from the tangle of riverine growth so their women could plant millet. Our arrival was an excuse to set aside their iron hoes. The young men wanted to ride our boats to the next settlement downstream, but their father forbade it. Lack of a common tongue frustrated our communication. Soon our oars dipped water as the primitive hoes of the Bodi gouged the earth in their seasonal struggle to raise a crop along the Omo.

As we approached our first major settlement of Bodi, every woman and child in sight dropped everything and fled to hide in or behind their small huts woven from limbs and grass. This region was so remote that only a few people here had ever seen a white person. At this time we were mostly a matter of legend. A few nonchalant warriors remained to meet us. These lithe, clean-limbed Nilotics had shaved heads and shaved pubic areas and wore nothing except for an occasional strip of cloth pounded from fig tree bark or muslin like trade cloth draped over one shoulder or wrapped around the waist as a cummerbund. They were armed with spears, daggers suspended down the back by a throng around the throat, and a few rifles (mostly Model 1895 Austrian Mausers) hanging from one shoulder by homemade slings. The flight of the females and children contrasted with the

casual wariness and show of ability of the adult males immediately struck me with their close parallels to the fright responses
of chimpanzees and baboons. I wondered how much of this
resemblance was due to instincts we shared with our nonhuman
primate relatives versus how much was due solely to human
cultural traditions of males as protectors. Soon the people of the
lower Omo would raise an even more important question concerning human sociobiology.

Reassured by our friendly hand-shaking with the men, one
small group of women and children after another cautiously
returned to study us. Their heads were also shaved, maybe to
combat lice. Hair styles of all ages differed individually, but
consisted of leaving small, symmetrical patches of scalp un-
shaved. One style popular among men was to let a single forelock
patch grow long, braid it, then tie off the end into a small cowrie
shell traded from the distant Indian Ocean. This forelock was
literally the only hair allowed to grow anywhere on the entire
body. Most women and some men had pierced ears with dis-
tended lobes encircling clay or wooden plugs. Bodi women also
wore small plugs of wood from a tree called *"bodi"* in their lower
lips, and by doing so gained the name by which other tribes know
them. They called themselves Ma'an. While this disfigurement
may sound ugly, the visual effect was minor, though it did cause
the women's speech to slur. As a vast exaggeration of this, most of
the women of the Mursi tribe we would meet downstream inserted large, Ubangi-style clay plates in their lower lips that
stretched them tight. According to one theory, these disfigure-
ments were a carryover from the days of slave raiding on the
upper White Nile in Sudan (from whence the Bodi and Mursi
emigrated several centuries ago), when women were purposefully
disfigured so as to reduce their desirability to the customers of
the slavers. It still worked.

Bodi women wore capes of goat hide draped over one shoulder
and fastened on the opposite hip, leaving one breast bare. Attached to the fringe along the margin of a cape were jangling
collections of ornaments fashioned from bits of bone and ivory,
bell-shaped pendants of hammered metal, glass and porcelain
beads, spent Mauser cartridges—anything, it seemed, that contrasted with the austerity of their lives. Most Bodi women appeared healthy and attractive. Children went naked, but had

multiple strands of red, white, and blue beads around their necks, waists, wrists, or ankles. Altogether the Bodi appeared primitive and barbaric, but for all that, their pride and dignity were even more apparent. Their style was still in harmony with the rest of their lives, as yet untainted by the insidious influence of status-conscious Western dress now oblivious of the environmental function of clothing or, even worse, by evangelical missionaries who regard the human body as a shameful exhibit.

Both women and children were camera shy, but the latter fell into paroxisms of glee when we showed them how to look through the viewfinders of the cameras. They convulsed at how small and funny their playmates looked inside these odd boxes we carried. Stevie Bruce treated the adults to samples of their own impromtu speeches, most of which were indecipherable to us, from her portable tape recorder. Suzie Yee handed out instant polaroid snapshots that riveted the adults; we probably could have traded one of them for an entire goat (although their cows and goats were far from the river and the tsetses). I felt like a tourist for the first time in my life. But we may as well have been from another planet for the hiatus in culture and technology between us.

Just before we left, Yost handed out gifts of Chinese bar soap while Charles dispensed chloroquin pills for malaria and tetracycline for infections. These commodities, along with razor blades and rifle cartridges, were the products of Western technology that enticed the Bodi most. Sobek policy on the Omo was not to give or trade to people *anything* that they could not already procure through some existing trade network. This was to avoid cultural pollution and an inadvertent unequal distribution of wealth. Emperor Haile Selassie apparently had made these old Mausers available to several pastoral tribes plagued by *shifta* (bandits), and via long-established trade networks, and at very high prices, these tribes ended up with a few of them. These days even cartridges were rare and expensive. The Bodi told us they sell their cattle to raise cash and pay seven Ethiopian birr (three American dollars) to distant officials for *each* cartridge. These cartridges are so valuable that they even substitute for livestock in payment of the traditional bride price: five cartridges equal a goat and twenty a cow.

During the rest of that day and most of the following we floated through Bodi territory, but by afternoon we had arrived in Batcha country. We camped on a sand bar where we usurped seven crocodiles basking in the sun. Thirty-two Batcha visited us as we set up camp. Only a few years before, these people, also known as the Kwegu, were unknown to anyone but their neighboring tribes. Their physical appearance was generally similar to that of the Bodi, even to the lip plugs, but as a group they did not seem as healthy or attractive, though they acted far friendlier.

Unlike the Bodi upstream and the Mursi downstream, the Batcha live along the Omo almost permanently and they hunt and gather to augment their crops of millet. Tsetses prevent them from keeping livestock near their settlements, consequently they have none, and, because of this, are looked upon by the other pastoral tribes as very low in the scale of humanity. The Batcha's lack of livestock has also led to a very bizarre dependency on the Bodi and Mursi documented by British anthropologist David Turton. Among nearly all African people, when a man wants to marry a woman he must pay a bride price to her father. Although difficult to understand, Batcha fathers now only consider livestock, or its equivalent in rifle cartridges, acceptable. To get livestock for a bride price the prospective groom now must solicit a wealthy sponsor from among either the Bodi or Mursi to provide cattle or goats from his more permanent living area and pasture several miles from the tsetse belt paralleling the river. Because even in Ethiopia there is no such thing as a free lunch, in exchange the Batcha groom becomes his patron's serf, providing ferry service across the Omo (neither Bodi nor Mursi construct or navigate the crude dugouts), plus many other considerations on demand which erode the Batcha's autonomy. It turns out that nearly every Batcha man "belongs" so closely to either a Bodi or a Mursi that the Batcha actually assumes much of the identity of his patron. And each Mursi or Bodi benefactor considers each man he assists as "his" Batcha.

These Batcha visiting us now studied our routine of setting up tents and cooking dinner as if we were skilled magicians from whom they hoped to learn an arcane art. We traded bar soap, razor blades, and rock salt for the small wooden stools carried by every man of substance. We passed beyond their country the next

139

day to enter deeper into the Plain of Death and the territory of the Mursi.

The Mursi were haughty and lacked the propriety of the Batcha, whom they consider fish-eating baboons. According to Naguce, a Mursi we would meet later that day who spoke Amharic, the Mursi, 3,000 strong, were at war with an unknown number of Bodi forty miles upstream. The hapless Batcha act as an unwilling buffer between the combatant tribes, both of whom emigrated to the Omo later. Batcha are held in contempt by the Mursi, who boast of kicking and beating Batcha men when they encounter them and of extorting a honey tax from the numerically inferior tribe.

As I learned more about the strife between the Bodi and Mursi, I saw more parallels between these warring tribes and aggressive communities of chimpanzees reported from Tanzania. The war between the Mursi and Bodi was primarily over territory: both tribes were growing in size; according to what I had read, in fact, outgrowing the territories they held. Both were pastoralists who required good grazing land and perennially watered sites for their permanent dwellings, but both commodities were very limited to the point where it probably would be fair to say that the Bodi and Mursi had reached the carrying capacity of their habitats (exactly as chimpanzees do) and that, to increase their prosperity and find space for their sons to carry on (which equates to increasing their reproductive success), their only option (other than adopting some new technology to make a living) was to wrest a chunk of territory belonging to alien males. Members of both tribes probably felt that they were doing the right thing, perhaps the only sensible thing, by warring on the other. Sociobiologically, the ecological and psychological rationales for war between the Bodi and Mursi or war between two communities of chimpanzees were basically identical, as was their inescapable dependence on solidarity and mutual support among themselves. Perhaps the only real question arising from this comparison, and it is an important one, is whether the warlike behavior of men and that of chimpanzees results strictly from cultural learning, or partly from a genetic component that predisposes the males of either species to war against alien males for personal, but also communal, gain.

Because we had six boatmen but only five boats, one of us had been taking time off the oars each day, and this morning, our last on the river, I had the first half of the day off. I rode in the stern of Slade's boat and kept an eye out for tsetses on his back.

Slade was rowing the lead boat. We had just inspected and rejected a wide mud bank as a potential lunch spot. As we drifted down a riffle where the Omo narrowed, we scanned the shore for a better spot. We had seen some huge crocodiles lately and I was hoping to get a good picture of one.

Like a thousand-pound arrow released from a giant bow, a thirteen-foot croc plunged into the river from the right bank and swam upstream directly at us. It came strongly and adjusted precisely to our changing position as we continued downstream. It looked like a crocodile-headed torpedo from some old World War II movie. Slade yelled "CROC!" and stood to hurl a croc rock, one of several fist-sized cobbles we stashed on board to discourage croc attacks.

As Slade hurled the first stone I deliberated over trying to get what might be a great photograph versus helping to stop this attack. Reluctantly I set the camera down and grabbed three cobbles. The last two of our croc rocks arced toward the saurian simultaneously, but missed. We were out of ammunition and the other boats were too far behind us to help.

As the final rock plunked into the water, the croc submerged. Slade grabbed the oars again and rowed downstream as if this might be his last race. A few seconds passed and the croc sur-- faced less than two boat lengths upstream. It eyed us briefly, then disappeared beneath the surface. Slade now rowed even harder. Stu, Mort, and Abby huddled together on the floor in the bow. Remaining on a side tube was unwise, because if the reptile bit into it, it would deflate and possibly dump the sitter into the river, a bad place to be.

Wickedly-toothed jaws eighteen inches long erupted from the river slightly more than three feet from the port side of the boat. Murky brown water flowed from that gaping maw as if a special-effects man had planned it that way. It was an apparition from the age of dinosaurs and it wanted us. I grabbed the camera and shot without focusing (the photo again turned out to be a blur, this one of jaws and white water). Conveniently the apparition had surfaced directly under the left oar. Slade slammed it down with a thud, then, encouraged by the solid feel of it, slammed it down again and again. The saurian submerged after the second thud.

Meanwhile, I fumbled to slice the tie on the spare oar, a fair weapon if the croc returned. Slade resumed rowing hard. A few seconds passed and the croc surfaced about fifty feet away to give us a final scrutiny, then vanish beneath the milk chocolate Omo. I slid the spare oar back in place. "Whew!" "Hey, Crocodile!"

We passed a half dozen parties of Mursi along the twenty miles of river that last day; their villages grew in size as we progressed downstream. Somehow the Omo felt different. Seventeen days were too few for this complex and isolated world. We were leaving too soon.

A solitary African buffalo raised his massive head above the waving grass at the water's edge. His wide-sweeping recurved horns gleamed jet black in the sunlight. Suspiciously he tested the air for our scent, then trotted off to a thicket. A few minutes later four lions looked up from the shore. With casual feline grace they climbed the steep bank, gave us a long last look, then melted into the narrow strip of bush behind.

These were wild lions, real lions, not the bored-looking, tourist-hounded beasts of the commercial national parks hundreds of miles south. Abruptly, the hysterical cry of an African fish eagle pierced the air, as if to remind us who owned this land.

This was Africa, the Africa whose primitive and unexplored vastness proved irresistible to men like Stanley, Livingstone, Burton, Speke, Thompson, and other intrepid adventurers of the last century, men for whom the challenge of this unexplored land was the song of a siren. And we were about to leave it behind. . . .

That afternoon we stopped at a Mursi village and met Naguce, a young warrior and friend of the Sobek crew. He climbed on one of our boats and continued with us to our camp and take-out point at the confluence with the Mui River. As I rowed John's boat, the sky blackened almost suddenly and thunder reverberated across the water. Torrential rain drove upstream so hard that Barnie, Seleshie, and Joe huddled on the floor to minimize their exposure to the bulletlike drops. Like the kind of dream where one runs mired in molasses, I struggled to row downstream against the wind on a stormy sea that moments before had been a placid river. Even to turn and face it to check our direction was punishing. Though ominous, the storm abated abruptly at the confluence.

While derigging the boats, we were careful not to step into the Mui because of bilharzia. John, Bob, Ward, Al, and Jim immediately packed their bivouac gear and started on the twenty-mile hike to the headquarters of Omo National Park to request their two Toyota Land Cruisers to pick us up as previously arranged.

Among the Mursi filing into camp was a somber woman with her ten-year-old son who had been bitten by a puff adder a month before. He had survived, but the tissue of his left arm medial to the elbow had rotted away. His eyes revealed he had resigned himself to death. Charles and Margie carefully cleaned his wound of dead flesh, which must have hurt, but the boy gave no sign. Charles explained to Seleshie, who explained to Naguce, who explained to the boy's mother what steps she must take to keep him alive. Rick, a doctor, thought his chances were nil without a skin graft. But if attitude can heal, the boy had a chance. His face was transformed to unbridled hope at having received the famed medicine of a white doctor. With a pang, I thought of my own son, Conan, to whom I had been a single father for so many years, and I wondered how he was doing back in the United States.

I had hoped to learn of the Mursi boy's condition from the next Omo expedition, scheduled in late October, but politics in Addis Ababa prohibited its departure. In 1978, politics were worse and no expedition ran the Omo. I did not return until 1982, but even then I was not able to learn whether he survived.

In the morning we found fresh crocodile tracks in the mud between our tightly packed tents. Both Toyotas arrived late that afternoon. Everyone piled in with their personal gear except Bart and I, who remained at the confluence to guard the equipment until the next day when the Land Cruisers could return. As they roared into the bush, the Mursi filed north, surreptitiously absconding the two bedraggled chickens they had traded to us earlier. At the Mui the Mursi women hitched their goat skin capes clear to their chins, and with much giggling and bouncing of lip plates, a single file of shapely derrieres vanished to reappear glistening on the far side.

Bart and I were alone with jars of peanut butter and jelly,

granola, powdered milk, a tent, and a heap of boats and gear. A lone vulture perched overhead and eyed us balefully until darkness fell.

Our company flew north from the Omo four days later in a chartered Ethiopian Airways C-47. The river seemed a giant brown artery feeding a drying land. I felt grateful for having been able to make this journey down the Omo and through time. Our visits with the Bodi, Batcha, and Mursi especially had given me a new perspective on the ecological and sociobiological concomitants of primitive war and the conditions that favored the development of warlike societies. They made me wonder again which factors were most important in spurring groups of related males to risk so much by attempting to wrest territory from other groups by pure force in what they *knew* was a potentially lethal contest. As the river shrank from my view, Kibale Forest immediately filled my mind. What was I missing? Suddenly I started to worry. In my mind's ear I could hear chimpanzees pant-hooting all the way from Ngogo.

—— 9 ——

It's a Jungle Out There

I managed to return to Ngogo in time to conduct my censuses and collect data on my phenology trees before October ended, but just barely. November brought heavy rains, the heaviest of 1977, and with the rains fruit appeared in relative abundance again. By now I was convinced heavy rainfall was a statistically reliable predictor that more fruit would be produced and that I would find more chimpanzees in the forest at Ngogo. Fruit was the weak link in the chain, but its presence alone did not guarantee its availability to the apes—fruit was sought avidly by dozens of other species of foragers, every one of which was inimical to each chimpanzee's pursuit of a full stomach.

One of my phenology *mucusos* in the south proved perfect for a new vigil. My first visitor was a chimpanzee I had named Spots due to the white patches in his beard. Like other old males, Spots seemed content to forage sedately and rest by himself. But his solitude was soon disrupted when Clovis arrived at the tree with her two sons, Clark and Chita. For the entire two hours the four of them spent in the tree, no move was made to socialize with Spots. It was as if he did not exist.

Chita had grown since I had first seen him nine months earlier, but Clovis still transported him everywhere in the crown. After she and her children abandoned the *mucuso*, Spots was soon joined by another male, whom I never saw clearly but whom I suspect was S.B. During their hour together, passing near to one another in the never-ending quest for figs was the closest thing they did to socializing.

But Spots' solipsistic bent was not unique in chimpanzee society; chimps often fed together for hours but gave only subtle hints that they were aware of one another. I probed this subtlety by analyzing all my vigil data from fruit trees and found that, statistically speaking, the apes *were* responding to one another quite significantly in ways that reduced the possibility of a confrontation or conflict over food. First, each newcomer arriving at a fruit tree usually chose to forage in the section of the crown furthest from chimpanzees already foraging. They were so good at this that each additional forager decreased the average distance between any two of them only by about a foot and a half. For instance, two apes foraging in a tree normally averaged thirty-five feet apart, but a dozen in the same tree still managed to remain an average of sixteen feet apart while foraging. And the bigger the tree, the greater the average separation: three chimpanzees in a small tree averaged less than twenty feet apart but in a large one nearly thirty-five feet.

Why were they so formal? Modern ecology recognizes that an individual's greatest competitor, by virtue of having exactly the same needs, is a member of its own species. So, unless they are as closely related as brothers such that each benefits genetically through the well-being of the other (through inclusive fitness), a chimpanzee's worst competitor is another chimpanzee. The apes' subtle efforts to give one another a wide berth while foraging, even in a crowded tree, keeps overt competition to a minimum. This automatic response probably is a vital adaptation to make their fusion-fission social system feasible. Without such propriety, usually on the part of the less dominant apes, there would be only fission, as among orangutans. The most fascinating aspect of this is that, once individuals have foraged and are satiated, they often close the gap between one another to zero and immerse themselves in mutual grooming. It does not take an observer long to learn the primary rule of chimpanzee deportment: feed first, socialize second.

One chimpanzee is another's worst competitor qualitatively, but quantitatively it is another story. For each ape at Ngogo, nearly 200 monkeys exist (see Chapter 2), most of them interested in the same fruit trees. By now at Ngogo I had learned two important facts about chimpanzees that I had little suspected, and that have either been given short shrift by other scientists or simply not observed by them: chimpanzees face severe competition for their preferred foods at times, and some of these competitors display such coordinated aggression that the apes end up losers. Some of these competitors also are preyed upon by the apes. Because I believe it is impossible to understand the lives of wild chimpanzees without considering the competition with which they must contend, this chapter describes the intricacies of their endless feud.

On a later morning at this same new *mucuso* a trio of young adult females arrived: Zira, Owl, plus a third whom I had glimpsed only once before but had not named. She was very distinctive: the upper margin of her left ear was missing a large chunk, and her right arm ended with a stump instead of a hand. Most likely she was a second surviving victim of camouflaged wire snares set in the forest by poachers (Chapters 10 and 11). She was a small female about Owl's size and age. So far my names for females had reached down the alphabet to "h" for Hump (a very old female); "i" was open and next in line, so this new female became Ita (pronounced "eeta").

Ita was an unbelievable daredevil. Chimpanzees are excellent climbers but also generally cautious. Despite being handicapped, Ita regularly leapt across empty space and lethal heights to move from one major limb to another, rather than return to the bole of the tree as did most apes. Her acrobatics sometimes held me in tense fascination; I often wanted to face away as I watched her flex her knees in preparation for a death-defying leap, so as not to witness her imminent suicide.

They foraged for two hours, then Owl and Ita settled into mutual grooming. Zira moved into an adjacent tree, a tall, thin *Diospyros* swaying in the wind, constructed a substantial day nest eighty feet above the ground, and settled into it for a half-hour nap. Wind rocked her back and forth in what seemed to me a radical arc, but she never poked her head up to reevaluate her choice of trees. Later the trio resumed low-key foraging for two more hours. Then Owl faced south, looked down, and pant-grunted an excited greeting females reserve for adult males.

Silently R.P. climbed up from the shade of the *mucuso*. He gave no overt sign of recognition toward Owl or either of the other two females, or to me. (So far in my observations, Owl had allowed only one male to mate with her, R.P.) While moving to a better site to observe Zira, my focal individual, I walked under him. As he glanced at me his left hand crept around his side to lightly touch and pat the healthy skin adjacent to the angry-looking wound on his back. It was obviously infected and a chronic source of irritation which he had learned was unwise to touch directly. Although I had seen this idiosyncratic habit of R.P.'s dozens of times by now, it never occurred to me that I might be able to do something about it. One of the implicit laws concerning a scientist's relationship with the animals he observes insists that he remain neutral. For most people who get to know their subjects as individuals it becomes very difficult not to start feeling fond of some of them and to want to alleviate their pain when aid is a simple matter. With creatures as manlike as chimpanzees, I think it is impossible not to actually start thinking of them as humanlike, at least subconsciously, and then to feel empathy. Such anthropomorphism is rightly the bugaboo of the behavioral sciences; however, with chimpanzees it is not only inescapable, but, when coupled with rigorous objectivity, it is an asset to understanding them. At any rate, this was when I started carrying a tube of antibiotic salve in my pack, just in case I got close enough to put it on his wound.

After foraging as if the females did not exist, R.P. scanned the variety of routes out of the *mucuso*, then chose the one leading to where I sat. Halfway down he paused and looked at me. I glanced away and scratched. He scratched too, with a sandpapery sound, then descended to the ground a few yards away and walked off. It was hard to believe he was the same ape who jumped out of a tree upon first seeing me; now he had come closer to me voluntarily than I ever expected of a wild chimpanzee. Salve.

The females continued eating as if this was their first opportunity. Where those figs were going was a mystery. I do not know how long they would have kept it up; they had been at it now for more than seven hours, but after R.P. left red colobus moved in from the north.

An adult male colobus entered the *mucuso* and vocalized in threat "chists" near Ita, who immediately moved five yards further away. After a few seconds the monkey rushed toward her. Ita screamed repeatedly while fleeing toward the east exit route. The colobus chasing her was joined by two other adult males from his group. Ita fled, but paused long enough to glance over her shoulder at the attacking monkeys. Seeing them in hot pursuit, she made reckless, headlong dives down through the foliage from tree to tree. I watched her wild one-handed acrobatics in disbelief. Zira screamed and rushed toward the chase to intercept them, but Ita almost reached the ground before Zira arrived. Without a pause both females disappeared, while three red colobus threatened them from low in the trees. I had not seen Owl descend, but when I looked for her she had vanished.

Soon I staked out a new *mucuso*. Most of its figs were still green but it had enough ripe ones to attract chimpanzees. This vigil was almost unique because it offered a dry and comfortable place to sit, the bole of a fallen giant that had been stripped of bark. I sat on it and gazed up at the *mucuso*. Then I lay back along it and still had an excellent view. The place was perfect. Too good to be true.

Red colobus drifted into the tree. While scanning the crown I realized that at least half the green figs that had been up there two weeks before were now missing. More red colobus arrived. More green figs vanished. The group swelled to about fifty and consumed unripe fruit like a Henry Ford assembly line.

For months I had been anticipating this next major crop of *mucuso* figs. I monitored each giant tree that provided me with past observations so I could predict within a week or so when the apes would arrive again for figs. I expected my current familiarity with the chimps and their tolerance for me to make this second round at the *mucusos* pay off big. In fact, I was counting on those trees like the go-for-broke gambler who *knows* his horse will win.

But on the eve of ripeness, the green figs of each *mucuso* had disappeared, as if supernaturally, without a sign of the apes. Now I was seeing why. Despite most of their diet consisting of leaves, red colobus had a taste for *mucuso* figs in a state that chimpanzees would not eat until they ripened a week longer, probably due to toxins. Because an average troop of red colobus contained about fifty monkeys, a two-hour visit by them resulted in the consumption of thousands of figs. It was as if they were on a communal green fig binge. Apart from their actual consumption, their wastage was phenomenal. Quite unlike the apes, these monkeys often picked an entire rubbery cluster of a dozen or more figs, ate one, then dropped the rest to be scavenged by bushpigs. It was almost as if red colobus were removing figs not simply to satisfy their hunger, but to make the tree, perhaps the entire region, less attractive to chimpanzees. As these figs vanished, I suspected so would my future opportunities to view apes from the perfect log, or in any *mucuso*.

The foliage rustled about fifty feet away. I sat up. Owl climbed about a dozen feet up a tall sapling leading into the *mucuso*. She glanced at me and hooed in surprise. Her mouth was jammed with something resembling mashed figs, probably fallen ones.

She gazed into the fig tree, now crawling with monkeys, and hesitated. She looked at me again and hooed, as if in disgust. Two adult male red colobus left the *mucuso*, climbed toward Owl, and perched directly above her to chist in mild threat. Doubtless their hostility would increase were she to climb higher. Owl studied them and hooed again.

One of the males spotted me, abruptly rushed several yards higher, turned, then chisted at me a bit more intensely than toward Owl. The monkeys in the *mucuso* paid them no attention; none of them had seen me despite the two hours I had been below them unconcealed on the log. This was a standoff: the males could not descend to attack Owl because I was near, and she could not ascend because they would never let her past them. I briefly considered scaring them away, as a favor from one hominoid to another (Hominoidea is the superfamily containing great apes and man), but of course I did not do it.

Owl perched patiently for more than an hour, but the red colobus would not budge. Finally, driven by hunger or impatience, she started stripping the new leaves from a nearby tree and eating them. The monkeys showed no sign that they would leave soon. Although both males had stopped branch-shaking and chisting, they continued to perch aggressively above her. Finally she descended and walked north, possibly to another tree she had in mind. Twenty minutes later, the colobus vacated the *mucuso*.

No foragers visited the tree in the next four hours, so I headed back toward camp. Only a quarter mile southeast of it I passed beneath a large *Pseudospondias*. From above came a low "uumm" I recognized. I looked up to see Owl gazing down at me from amid clusters of olivelike fruits. She was not eating, merely sitting.

After finding a nearby spot with a clear view, I sat down and observed her askance. She lolled in one comfortable crotch after another for an hour, alternately grooming herself and gazing into the sky. A group of redtails approached from the west. Owl sat up abruptly, uttered three deep "uumm's," moved about ten feet to a cluster of fruit, and started eating as if under the onslaught of a terrible hunger. Five minutes later the vanguard of about thirty monkeys entered her tree and began picking fruit. They stripped and discarded the pulp and stored the long seeds in their bulging cheek pouches.

More redtails entered the tree. Ten minutes after she started feeding, Owl exploded with a "hoo-grunt" and feinted a lunge toward an immature redtail foraging about twenty feet away. The young monkey glanced at her, then leapt completely out of the *Pseudospondias* into a sapling below. It was impossible for her to chase them all out and monopolize the tree. Being only a tenth her weight they were far more agile and could simply maneuver around her in circles.

Five minutes later Owl climbed down the bole of the *Pseudo-spondias*, then paused at its base to study me for about fifteen seconds, a long look. Then she slowly walked in the direction opposite me toward a clear little creek gurgling down the hillside. She paused again at the creek's edge to gaze at me briefly. Then she squatted on all fours to suck up creek water for a few seconds. She glanced at me again, took a quick drink, stood up, crossed the creek, and casually walked into the gathering dusk without another backward glance. She trusted me, but not completely. As it had so many times before, the worn greeting from some space drama formed in my mind and became stuck because I had no way to say it, "I, an alien to your world, come in peace. . . ."

It was hardly surprising that monkeys and chimpanzees foraging for the same foods would clash, but it took me quite a while to size up the magnitude of their conflicts. Monkeys, other than red colobus, often deferred to male chimpanzees by waiting until the latter had vacated a food tree before entering it—or else they foraged in it warily. Solitary male monkeys were more casual about the apes, and sometimes even groups of monkeys joined adult female chimps with their offspring. Despite what seemed to be at times a superabundance of fruit, and despite the monkeys' caution, mere proximity between apes and monkeys sometimes exploded into dynamic conflicts. Because every clash between them was both a conflict between *competitors* as well as strategic maneuvering between *predator* and *prey*, I found it hard to interpret some episodes, like the following.

Many redtails and mangabeys had been foraging together but now rushed in a wave out of a gigantic *Ficus dawei* to linger in a swamp nearby. Soon after their exit, six chimpanzees followed them. As I looked to see if R.P., the only adult male present, was among them, both species of monkeys chorused simultaneous, high-intensity alarm calls. I craned my neck to see an eagle glide down the swamp corridor away from the monkeys. Was this the danger? During the previous eagle attack I had seen at this *dawei*, when the bird had dived into the crown alarm calls had been immediate and monkeys had simply dropped off limbs into space. They had looked for ways to arrest their falls en route downward.

This *dawei* was relatively isolated from other big trees and offered a clear flight path for eagles. But even though the eagle was gone the alarm calls redoubled. Redtails and mangabeys streamed away from a nearby low tree that vibrated as if from a vigorous struggle. Frustratingly, I could not see what was happening.

A chimpanzee screamed. Then I heard animals scrambling toward me through the undergrowth. I moved closer. A large juvenile chimp galloped so fast toward me that I did not recognize him. Only a few yards behind him was an adult male mangabey in hot pursuit. The chimpanzee was oblivious of me only a dozen feet away but paused to look over his shoulder at the mangabey. He then hooked one arm around a sapling and swung upward to perch on a horizontal log across the trail. The monkey skidded to a halt at the ape's feet. The young chimp bent down to grapple with him. Then he leapt to the ground and fled southwest. The mangabey ran west, then climbed a tree to return to his group.

Although incomplete, I could not help trying to interpret this episode. My guess was that the juvenile ape may have taken advantage of the monkeys all having dropped into low vegetation in response to the eagle to attempt to nab a young one. S.B. Group of mangabeys contained a one-year old; the redtail harem had several infants. Apparently at least one adult male mangabey interrupted the attempt and routed one ape. Were any of those infants now missing? Soon the mangabeys moved to the edge of the swamp where I counted all fourteen of them. The redtails moved further away and, because I did not know how many infants had been in the group, I was unable to determine if one was missing. Was I right or wrong? Had an ape grabbed a monkey? Or even tried? I could only guess, as I was forced to do in the next, even more dynamic example.

While walking a section of trail I heard a single scream of a young chimpanzee about two hundred yards north. I continued north about half that distance and slipped unnoticed beneath several redtails in a large tree. A second scream helped me zero in twenty-five yards through the understory to see an adult male chimpanzee, whose face was turned away, calmly walk away from me along a fallen log. A juvenile sat behind him and gazed about furtively as if to check whether anyone was watching him. He did not see me. After a minute or so he followed the adult.

As I looked for a good route to follow them, red colobus alarm calls erupted from less than 100 yards beyond the log. As these calls intensified I deliberated over whether to sneak in among them and risk interrupting their interactions, or remain and possibly be able to investigate their outcome. Then I heard animals rushing toward me from the fracas. Both chimps appeared. The large male veered south running fast, while the juvenile disappeared to the west. Behind him an adult male red colobus halted his pursuit and climbed up from the ground into a tree. Moments later more chists, then the adult male chimpanzee burst from the understory to race east again in reverse along his original path of flight. About ten yards behind him two adult male red colobus pursued him hotly along the ground. All three were engulfed by the foliage. For the next few minutes I searched the area in hopes of finding more clues to the puzzle, but I had no luck. As if a message of defiance by the apes, frenetic drumming of a tree buttress reverberated sonorously from nearby in the north. Then it sounded like the red colobus went crazy.

Here again I could not avoid suspecting that I had just witnessed part of an attempt to prey on monkeys that failed when the adult males defended their troop by chasing the apes on their own ground. The number of similar incidents at Ngogo I observed in part prompted me to wonder how many others I was missing, and how many resulted in the apes killing a monkey. It was difficult to believe the apes would keep trying, and the monkeys put up such stiff defense, if the apes never succeeded.

Chimpanzees at Gombe attacked, killed, and ate red colobus and other monkeys. Jane Goodall (1986), David Riss and Curt Busse (1977), Geza Teleki (1972), Richard Wrangham (1975), and others reported numerous incidents of hunting, almost all opportunistic and nearly *half* successful. Curt Busse reported in 1977, for instance, that between 1973 and 1974, the Kasakela Community killed thirty-nine red colobus at a rate of 0.95 kills per hundred hours of observation.

But during thousands of hours of observation of red colobus by Tom and Lysa, they saw no successful predatory episode, although they did witness three probable attempts. I suspect their observations and mine reflect a real difference between hunting by chimpanzees at Gombe and Kibale, and I think the difference is due less to the apes than to the types of habitats in each region. Wrangham (1975) suggested that monkeys at Gombe were most vulnerable to chimpanzees when in broken canopy or low growth. Tim Clutton-Brock, who studied red colobus at Gombe, and Curt Busse reported that red colobus frequently chased chimps away during their unsuccessful attempts, especially in high canopy forest. Clutton-Brock even saw one chase on the ground similar to mine described above. Notably, Clutton-Brock never saw chimps catch a monkey from his Gombe study group.

Much of the habitat at Gombe is broken canopy or low growth; relatively little of it is tall forest typical of Ngogo. Chimpanzees are excellent climbers but are cautious when high above the forest floor. Their ability and mobility is inferior to that of the much lighter colobus, who can easily outmaneuver them. The consequences of falling are far more serious for the apes—Jane

Goodall (1983, 1986) and Geza Teleki (1973a) reported one male who died this way and Goodall mentioned several other males (ten to one over females) injured as a result of falls. Monkeys also occasionally fall and are killed, yet some monkeys smack the ground but wobble to their feet apparently unscathed like Wile E. Coyote. Red colobus probably are killed by apes more frequently at Gombe because they must frequent habitats which make them more vulnerable when searching for food. Chimpanzees also preyed on redtails at Gombe, but less so than on colobus, possibly because the smaller monkeys tended to remain in forest habitats providing greater immunity. Eventually I concluded that, because the trail-grid system at Ngogo contains little habitat where chimpanzees normally could hope to catch a colobus, my opportunities to witness success in hunting were fewer than at Gombe. (As an example of how successfully some of the Gombe chimps hunt, one adult male at Gombe, Goblin, was seen to tear two infants from their mothers in rapid succession, present one each to his mother and sister, then rush back into the colobus troop to grab a third infant for himself!) But despite less favorable terrain the chimpanzees at Ngogo kept trying.

If red colobus increased their vulnerability by foraging in low growth, they offest this increase somewhat by living in multimale groups. Curt Busse proposed that the multimale group structure of red colobus was a social adaptation specifically to combat predation by chimpanzees. Almost all other small monkeys in Kibale, the l'Hoest's, blues, redtails, and black and white colobus, however, lived in one-male harems, yet seemed no more vulnerable than red colobus—in fact, they seemed less so. And both l'Hoest's and black and white colobus often used terrestrial or understory habitats which should have increased their vulnerability.

Although the social system of red colobus did seem to discourage monkey-nabbing by the apes, particularly in circumstances wherein the initial attempt failed, I suspect their multimale society was originally an adaptation of a different order. The sociobiological hint that this may be so consists of Struhsaker's observations that male red colobus rarely leave their natal troops, yet females commonly emigrate to be replaced by immigrants from other troops. This system, though extremely rare (among 181 species of nonhuman primates it has also been seen only among chimps, mountain gorillas, hamadryas baboons, and spider monkeys, though neither monkey is territorial) is identical to that of chimpanzees and can easily lead, through the mechanism of inclusive fitness, to cooperative defense by red colobus males of the troop's infants. The range of a red colobus troop is most likely a patrilineal inheritance, and, while not their communal foraging area from competing troops of colobus (and possibly from chimpanzees as well). The larger a troop, the more successful it is in elbowing aside other troops. The ability to parry predatory attempts by chimpanzees may be simply a serendipitous by-product of this kin-selected male reproductive strategy of red colobus.

But no matter what evolutionary pressures favored a male-retentive society among red colobus, they used this adaptation to the hilt when confronting chimpanzees. Close family ties between red colobus males may be critical in enhancing their cooperation and mutual support during dangerous confrontations with the apes. Significantly, the males from other multimale species, baboons and mangabeys whose adult males were all immigrants far less likely to be closely related, did not cooperate to drive chimpanzees from food trees despite being equally formidable individually and having just as much at stake—and they consistently ended up losing out to the apes because of this. At this point I had seen several females chased from fruit trees by red colobus males, plus two episodes of male apes being chased along the ground after what I suspected had been attempts by them to prey on colobus. It did not surprise me that male red colobus would risk battling adult male chimps to stop them from preying on one of their infants, but I doubted they would attack simply to reduce competition for figs. But, as the following episode from one of my later censuses shows clearly, the two motivations were difficult to separate.

I arrived at a fruiting *mucuso* in or near which were foraging five species: mangabeys, redtails, blue monkeys, red colobus, and five chimpanzees: Stump, Fearless, an old adult female, a young female, plus a juvenile male. The mangabeys and redtails were peacefully sharing the *mucuso* with the apes. Oblivious of me, red colobus filed one by one into the fig tree from the north. Immediately all five apes shifted to the south side of the crown; the young female was so frightened that she actually screamed and leapt down into an access tree. In one minute nineteen red colobus entered the crown. An adult male rushed toward the old female chimp, who stood her ground for a few seconds to grapple with the monkey, then she broke away by leaping at least ten feet downward into an understory tree. Within seconds all five apes had vacated the tree, but they remained near it and screamed sharply in protest. One-handed Stump roared and barked threats, and the colobus chisted sharply in response. The apes moved to another, nonfruiting *mucuso* about fifty yards away where old Stump continued his impressive but unsuccessful roaring and was accompanied by screams of solidarity from the other apes that again elicited instant choruses of threat vocalizations and branch-shaking by the red colobus, who, despite all the noisy protest, had usurped the chimps at their food source. Interestingly, the colobus did not displace either of the other two species of primates.

The upshot here is that during all fourteen clear-cut interactions I observed between them, the red colobus either prevented the apes from entering a fruit tree or caused them to flee one, and, in only about half the incidents, usurped them as well. It seemed a rather poor showing for killer apes.

Chimpanzees are not the only great apes to be dominated by much smaller primates. Herman Rijksen (1978) and John MacKinnon (1974b) reported that male siamangs (a lesser ape a quarter the size of an orangutan) launched coordinated attacks to drive Sumatran orangutans from fruit trees—strictly in competition for the fruit. Red colobus, however, exhibited a vengeance suggesting chimpanzees were far more threatening than mere competitors. Finally, I could not help but conclude that denying the apes their favorite food was probably not competition, except perhaps secondarily; it was a strategy to foil predators. By eating green *mucuso* figs the monkeys removed a prime incentive for the apes to visit their range and perhaps also drastically reduced the time spent by the apes in their neighborhood and, with it, the possibility of an opportunistic hunt.

Because the red colobus were killing two birds with the same stone, it would be very difficult to determine their motivation conclusively. The intensity of competition involving direct clashes between any two species is determined by two main factors: the overlap of defensible resources they use in common and the relative biomass of the competing species. I observed red colobus eating only 20 percent of the same foods as the apes, but the monkeys outnumbered them sixty-five to one and collectively outmassed them ten to one. In fact, red colobus accounted for 38 percent of the entire primate biomass at Ngogo (8,214 pounds per square mile). Baboons, redtails, and mangabeys made up much of the rest, and their diets overlapped much more than that of the red colobus with chimps, but collectively they clashed with the apes only about as often as the red colobus did on their own; also they usually came out losers. Clashes between red colobus and chimps seemed less like competition than a vendetta.

Other monkeys, however, were serious competitors. Based on the proportion of fruit in their diet, the apes were Ngogo's primary fruit specialists. But even where fruit appeared abundant, it was limited. By afternoon, for instance, most ripe fruit was gone, and foragers found less to eat even in heavily laden trees. Monkeys seemed to have the most impact of all competitors, but they were opportunistic: when a tree in a group's range bore fruit, they visited it on most but not all days. They also were more catholic in their tastes. Leaves, insects, bark, blossoms, and buds (foods also eaten sometimes by the apes) often outbulked fruit in their diets. Redtails overlapped the chimp's diet by 48 percent, mangabeys by 38 percent, and baboons by 24 percent. When a fruit tree was small or the apes numerous, they usually tried to chase all monkeys, squirrels, and fruit-eating birds away from it. The apes of Ngogo foraged over the home ranges of thousands of monkeys who competed with them. Just how important is this?

The concept of competition seems simple and unambiguous and all of us have probably seen examples of it we would swear were valid, but among ecologists serious contention exists concerning the impacts of competition between species—and even over whether competition occurs at all, and, if so, what constitutes good evidence for it. One of the canons of ecology, formulated in 1934 by G. F. Gause, is the Competitive Exclusion Principle, which maintains that when two species depend on the same vital but limited resource, one eventually will outcompete the other to the point of the second one's local extinction. By definition, of course, any two species found in the same habitat cannot fit into this category unless they have just recently come together. But nature is rarely this black and white, and many examples are known of two or more species vying to harvest limited, important resources. The very difficult trick is to determine whether they are competing to the extent that any of the species in the "arena" are experiencing a ceiling or reduction in their population due to the others. This would constitute *competition*. Part of this chore is determining whether each resource in question is a *limiting resource* which constrains population size. Because animals are often adaptable enough to use alternate resources when their first choice is taken, limiting resources are difficult to identify, and hence competition even more difficult to demonstrate, except in very simple ecosystems.

Despite hot debates over what constitutes unimpeachable evidence for competition, competition with fruit-eating bats and birds still takes the blame for closing certain avenues of primate evolution and, in fact, for causing several primate extinctions. The contemporary logic for competition is this: if individuals of a species suffer inadequate nutrition (especially likely during seasonal shortages) because members of other species are harvesting the limited resources that the first species needs, the natural consequences (that have been observed among baboons and Japanese macaques due to lowered nutrition) are poorer health, delayed maturation, shorter longevity, and the logical proof of the pudding, lower reproductive success. Documenting the proof of the pudding among large mammals requires a prohibitive investment by the investigator, though it has been accomplished experimentally with rodents. But there are ways to look at feeding dynamics and logically infer competition.

Tom Struhsaker (1978) compared the five most common food species of red colobus, black and white colobus, redtails, blues, and mangabeys at Kanyawara and found little overlap between them except for blue monkeys, whose diets overlapped the others most. But did they depress the population of blues? Struhsaker compared their population density at Kanyawara with those recorded in other forests where mangabeys and red colobus did not exist and found blue monkeys to be more abundant in them. This looked like good evidence for competition. To take it a step further, at Ngogo, where mangabeys are 30 percent more abundant than at Kanyawara, blue monkeys are *seven times* less abundant— being represented by only a single one-male harem in the entire study area! Is this Gause's Competitive Exclusion Principle in action? Or is it due to some other factor no one had guessed? At this point I was convinced that most species of monkeys at Ngogo were competing with the chimpanzees. We know that the vast difference in adult size and rates of maturation between captive and wild chimpanzees is due to better nutrition among captive apes; this suggests that any reduction in nutrition among wild apes due to other primates *is* competition. The mystery remaining consists of the degree to which they were competing.

To be successful against this horde of competitors outnumbering them 197:1 demands two superior talents: intelligence and the ability to dominate. Intelligence is a slippery concept because it is impossible to determine the absolute intelligence of any creature, humans included, without also determining the subject's motivation. The great apes are notorious in captivity for being bored with the tests of psychologists. But, as I mentioned in Chapters 3 and 4 (see also Chapter 15) chimpanzees are capable of amazing mental performances when food is the incentive. Duane Rumbaugh conducted an exhaustive review of learning skills and in 1970 concluded that the complex "learning set capacities of the great apes range to higher levels than any of those obtained for other nonhuman primate forms." Superior mental ability—for remembering locations and timing of foods and for planning efficient itineraries to harvest them—seems not merely an advantage but a necessity for chimpanzees to live predominantly on fruit in the primate-eat-fruit jungle. Being able to dominate most of their competitors also seems indispensable. The apes used every competitive advantage, including starting their foraging for fruit at dawn, earlier than monkeys. Still, interpreting many of the clashes between the apes and monkeys of Ngogo is difficult because the latter are both competitors and prey—hence the intensity of the permanent war between chimpanzees and red colobus.

For the answer to one question asked by these clashes, I would have to wait until nearly my last day at Ngogo. After nearly 500 hours of observing chimps in Kibale, Eskimo finally revealed a critical piece of evidence. Macho as usual, Eskimo gave the air of an entity to be reckoned serious by anyone. And, as usual, he was stuffing himself with figs. When he stood to move to a new cluster of figs this day, he was carrying a skin large enough to account for most of a young, red colobus monkey.

As he sat to eat figs, he alternated them with nibbles of meat from the inside of the skin. Forty-eight hours later I saw Eskimo again (and for the last time) *still* carrying his shrinking prize. I do not know how he got it, but, because chimpanzees have never been seen to take opportunities to scavenge animals found dead, almost certainly Eskimo's prize came from a kill by a chimpanzee, probably by Eskimo himself. But no matter which killer ape was responsible, the red colobus obviously had reason to harass the apes of Ngogo.

—— 10 ——

Cultured Chimpanzees

"*Habari gani* (what news)?" I asked while pumping Tom's hand. I had driven around Fort Portal looking for him and had now finally found him descending the front steps of the post office. Only twelve days ago Tom and Lysa had returned after being absent from Uganda one hundred days on a trip to the United States. A few days ago Tom had sent me a note via Mubu, his game guard, to rendezvous with him in early December in town so I could join him and Lysa on a trip north to Kabalega National Park. There, he and I were scheduled to give lectures on rain forest convervation and chimpanzee ecology at a conference organized by the Wildlife Clubs of Uganda.

"*Mzuri* (good)," Tom grinned, while juggling an armload of mail. "Well, a bit of bad news actually. Lukyamuzi has postponed the conference for a week."

"Any chance of buying enough petrol for a round-trip back to Ngogo in the meantime?"

"I'm afraid not. Even with this letter from the District Commissioner's Office, I can't get a thing. Fort Portal is dry." Tom added, with a barely perceptible sigh revealing his effort at getting over another hump in his reentry into Amin's Uganda. "I have to hold onto to everything I have to get us to Kabalega and back. I'm afraid you're our house guest, old boy. But the good news is the chimps have been shouting in K30 and K14."

"The chimps are around, eh?" I was glad to hear this, but already found myself anticipating Lysa's cooking, a second bonus. "I have some good news myself. Linda Scott recommended that I try asking the Boise Fund of Oxford University for grant support. Here's their response." I pulled the letter from my shirt pocket to show him. Funding was becoming much more of an issue now. Tom's grant ended this month, and my G.I. Bill award had already expired—by now I had been turned down for support a dozen times. We moved to one side of the busy stairs as I unfolded the letter and pointed out the critical sentence written by G. Ainsworth Harrison: "I am pleased to be able to tell you that the trustees of the Boise Fund today agreed to make you a grant of £500 towards the cost of completing your field study of the socioecology of chimpanzees in Kibale Forest Reserve, Uganda."

"Your first grant . . . congratulations!" He shook my hand again. "Took them a while, didn't it?"

Actually the support was welcome, but what impressed me most was the vote of confidence of the trustees for me, a stranger. Someone out there finally believed I could do it.

Wild buttress drummings and pant-hoots pierced the air repeatedly as if powered by a chain reaction. I couldn't help but smile as they drew me downslope through the forest in Tom's main study area of K30 (the Forestry Department's code name for this section). As I approached nearer I slowed and cautiously threaded my way along Kanyawara's unfamiliar trails. A large group of chimpanzees was creating this din and, if I was careful enough, they might prove to be a wealth of information. This was not my first experience with the chimpanzees of Kanyawara, but almost my first. Exactly six months earlier, in June, I had spent a week here counting their nests for a census and searching for the apes themselves. On only one day of that week had I hit the jackpot in a way that thoroughly intrigued me.

I had been drawn then by pant-hooting to a grove of *Uvariopsis* and had spent twenty minutes inching my way into it as slyly as possible. I knew if these apes were as unhabituated as those at Ngogo, I would cause a mass evacuation. One after another I had located eleven chimpanzees: three adult females each with an infant, plus four juveniles, and a young adult female with no infant.

One mother finally saw me. She climbed higher, hooed, then shrieked. I expected pandemonium to break loose, but instead, three juveniles stopped what they were doing, climbed down, moved closer (about a dozen yards away), and studied me intently. The other three adults paused in their foraging, glanced at me briefly, then resumed plucking the sweet red *Uvariopsis* fruits. The one nervous female burst into a medley of screams and shrieks, then climbed down and scuttled away. A large juvenile, possibly her older offspring, followed. Meanwhile, the other juveniles moved even closer and fixed their quiet gazes on me. I began feeling self-conscious, as if I were expected to do something.

Almost an hour after I entered the grove, two young adult females and a subadult male rushed wildly into it, pant-grunting as they came. Upon seeing me in its middle, the trio smoothly ascended trees without losing stride and began to feed.

A few minutes later nonstop pant-hooting and spine-tingling shrieks erupted from the wake of the three newcomers and half convinced me chimpanzees were killing some creature in an orgiastic bloodletting. As I moved closer to investigate, three adult males rushed out of the foliage and into the grove directly at me. Undaunted by my presence, they veered to one side of me, climbed trees, then screamed a new chorus.

Now I was in the midst of fourteen chimpanzees engaged primarily in feeding and secondarily in watching me. From such close scrutiny it was obvious that none of these was an Ngogo chimpanzee I knew. Despite being a stranger, they accepted me. Why?

I thought it was because Tom had been working in this area for seven years, although not on chimpanzees. And the Kanyawara Forestry Station had been here for much longer than Tom. North of here Forestry workers had cut down much of the forest, selectively poisoned parts of it to favor species they wanted, and replanted other sections with fast-growing exotics. Possibly this peninsula of primary forest jutting into Kanyawara remained attractive seasonally to the community of chimpanzees that traditionally had used the region.

Yongili had lived here before coming to Ngogo with Bill Freeland. He told me some of the workers here had been targets of aggressive displays by chimpanzees, who climbed down to rush along the ground dragging branches and pant-hooting toward them. Such displays indicate long familiarity. These stories also made me wonder whether the apes understood that the workers were destroying their natural habitat and their future. The men shouted and waved their *pangas* in the air to scare them off.

Another incident commonly related by residents of Kanyawara told how a Batoro woman walking along the road in the forest with a basket on her head saw a male chimpanzee suddenly appear in the understory off to one side. The ape commanded the woman to "Come" (presumably in Ratoro). The hapless woman put down her basket and followed the ape, only to be raped by him in the dense foliage.

Whatever the nature of their repeated contacts with humans, the Kanyawara chimps were nonchalant about me. One of the infants even maneuvered at her mother's teat, so that she could suckle and observe me without inconvenience. Eventually a greater number of adults simply sat around and observed me. A half hour of this and my novelty paled. Several adults, including two of the males who had climbed to the ground to sit on a log, leaned back to rest sleepily, paying me no further attention.

During the hour after their arrival pant-hooting in the forest finally ceased. This silence contrasted sharply with the prolonged din of earlier when chimps here now had been in separate parties, and it underlined the function of pant-hooting as a means by which those parties had stayed in contact.

Three hours after I entered the grove, the apes moved southwest beyond the trail-grid system of K30 toward Ngogo. I followed but eventually was balked by the understory. As the last chimpanzee vanished, I felt as though I had prematurely lost new but important friends.

Just how important was unclear. Their nonchalance about me made them a potential gold mine of information. But what kind of information? Because part of this community's range was the forest north of Kanyawara modified by felling, arboriciding, and planting exotics, their ecology was very disturbed, and, from a purist's point of view, so also might be their behavior. It was this problem, in fact, which had dictated that the Ngogo Community be my primary focus of research, despite their complete wildness. But the Kanyawara Community did offer me a contrast, a second opinion, as it were, one which intrigued me very much. Already their behavior seemed different—not just toward me, but toward one another. They seemed more socially cohesive. And despite the threads of conformity among chimpanzees, their idiosyncratic differences in personalities and behaviors increasingly impressed me. Did the Kanyawara Community really differ from that at Ngogo? These apes offered me the key to this mystery.

Their offer was metaphorical. After losing them in the understory I spent nearly a week trying to find them again in the small trail-grid system in K30, only one-third the size of that at Ngogo, which itself was too small to study chimpanzees and constantly had to be expanded. In fact, the Kanyawara Community seemed to spend very little time in Kanyawara and most of it to the south in trackless forest. So, unless I was willing to sacrifice time and resources at Ngogo, I could not tackle unravelling the peregrinations of the apes to the north. I was forced by circumstances to wait until they were again present in Struhsaker's study area before contacting them. And now they *were* here. . . .

Louder screams erupted from the confusion of forest ahead. Oddly, I soon found a quiet female plus her infant and juvenile daughters gorging on the fruits of a large *Mimusops* tree, a common species that had refused to produce good crops at Ngogo. For nearly three hours, as a bedlam of shrieks, screams, and hoots moved ever closer from behind me, I observed this peaceful scene. The females ignored me completely and continued to groom one another, forage, and rest together.

A noisy wave of nine chimpanzees of mixed ages and sexes finally arrived, paused to examine me, pant-hoot and display at me, or both, depending on the individual. Then they climbed and clambered above and around me to join the family in the *Mimusops*.

One young juvenile lagged behind, carefully balancing a hugely swollen, greenish foot too big for her body. Shocked, I examined it through my field glasses. It resembled a loaf of bread dough constricted in the middle by a tight rubber band. From one side of her gangrenous left foot projected the frayed end of a spiral weave cable, a poacher's snare. The rest of the snare was buried in her rotting foot. My stomach tightened and I felt an overwhelming empathy for her. I wracked my mind for some way to help her, but we had no tranquilizing gun, no veterinarian, no injectable antibiotics. Her foot looked beyond recovery, but I did not assume she was doomed; I had gained a respect for the powers of recuperation of wild chimps. There was nothing I could do.

She paused to examine her foot, then bit at the cable, but touching it was apparently too painful for her to continue. She hopped after her mother, who was carrying a small infant. Rage filled me. A poacher in my hands would not have lived long.

After foraging for *Mimusops* fruit for an hour and a half, a young subadult male descended to seven feet above the ground to sit about twenty-five feet away and observe me. He seemed fascinated, a response I was not used to. I felt an urge to move closer to him and somehow communicate my friendly intentions, but the hard-core scientist within rooted me to the ground. We covertly glanced at one another while I feverishly worried that I was missing an opportunity to somehow establish my trustworthiness, but I could think of no shortcut. Fifteen minutes later all the other apes headed southeast. The subadult and I followed them.

The group had absorbed the family I had been observing, plus four other latecomers, and now numbered sixteen apes. They congregated in a large *Ficus brachylepis* about two hundred yards from the *Mimusops*. This party was almost exactly the same size and composition as the one I had seen earlier in June here, and I thought that I recognized several individuals, but too many months had passed to be sure.

I settled near the fig tree to observe their efforts at defruiting. Rain soon prompted most of them to abandon the *brachylepis* and sit in the denser forest near me. Three juveniles and the subadult male approached to within twenty-five feet again and studied me. As before, I felt on the spot and self-conscious and wished that I could communicate with them. Between my data entries and writing and drawing descriptions for identifying them in the future, I pulled a Jew's harp from my pack and tried a tune on them. But they shifted uneasily at my twanging, so I put it away.

After an hour of rain, the party traveled together about another hundred yards northwest to a fruiting *Cordia*. They seemed less interested in feeding than in socializing; grooming clusters formed almost immediately. My subadult friend approached a hulking male, who stood, swayed to each side, then rushed toward him. The subadult leapt down about ten feet through the foliage and screamed for almost ten seconds. The awkward years. Though it may sound trite, they actually were; subadult males were the most common targets of the dominance actions of adult males, who often needed little excuse.

A female carrying a large infant three or four years old approached a second male but, before passing, stopped and fully extended her arm toward him. Almost absentmindedly, he reached about half the distance she had reached to briefly touch her hand. Thus reassured, the mother passed him to initiate grooming with another mother cradling a small infant daughter. The two infants grappled and mock-bit at one another as their mothers groomed, though the second female would not allow her daughter to stray from her grasp.

For an hour the *Cordia* was a grooming hot spot: the females groomed one another, the adult males groomed each other, and the females groomed the males. Even the subadult was groomed. Their lip-smacking pops were clearly audible where I stood scribbling notes on my data sheets amid swamp vegetation nearby. Shortly before sundown, nearly nine hours after I had joined them, they evacuated the *Cordia* en masse and continued traveling northeast beyond the trail system into K14. The last chimp down the *Cordia* was the small female with the snare embedded in her foot. She descended slowly and with exaggerated care, but even so, her foot wobbled painfully with each step. The group, including her mother, was leaving her behind, but like a trooper she was silent. I felt a desperate ache of empathy and cursed my inability to help her.

During the next five days before the Wildlife Clubs of Uganda Conference I again counted nests and searched for the apes. Exactly like my week here in June, however, I never found them again; they had vacated the study area. But even though I had literally spent only two days with them, I was now convinced that the Kanyawara Community would provide a valuable contrast to the apes of Ngogo. Their behavior among themselves was different, and I suspected some of the differences might be cultural. But due to the vicissitudes of research, years were to pass before I saw them again.

The next two-thirds of this chapter depart from the general chronology of this book; they occurred after nearly everything else in it. But they form the logical conclusion of this chapter *and* present critical observations that enhance and expand the chapters following. I returned to Uganda in January 1981 to begin a major investigation of the population status of chimpanzees in all of Uganda's forests. Dr. Milton Obote had just been reinstated as president many months after the liberation of Uganda by Tanzanian invasion forces. Uganda had literally been through hell, and through three other presidents who had lasted only long enough to arrange their furniture. Just one of the current symptoms was months of lack of petrol in the west, a shortage which again confined me to Kanyawara and Ngogo. This turned out to be a blessing in disguise.

My searches in Struhsaker's K30 study area were unproductive. But Joe Skorupa and Lynn Isbell, graduate students of U.C., Davis now studying red colobus in disturbed habitats, told me they had been seeing chimps north in K14, an adjoining area selectively felled during the years of the Uganda Protectorate. Although the understory was dense with regrowth, K14 was much more pleasant than the constantly up or down K30. A quick look told me what had drawn the apes: K14 was dotted with large *Ficus natalensis* trees, four of which were bearing fruit.

One of these held two apes: a large subadult male and a middle-aged male whose left hand was missing two and a half fingers; those remaining were frozen into spindly useless claws. My scheme for naming the Kanyawara Community was to include a "k" in each one's name, and, of course, to give a female's offspring names beginning with the same first letter as her own. The older male became Hook and the subadult Knuck, for a slight deformation of the hand.

Startled when he first saw me, Knuck pant-screamed, rushed to Hook, and fully extended his right arm to the older male. Hook casually reached to touch Knuck's hand briefly, and this contact immediately calmed him. Oddly, Knuck became utterly blasé about me in a day or two, and, over the next few weeks, increasingly curious and attracted to me—a déjà vu sensation. As I studied him I realized he was about three years older than the young subadult male who had been so fascinated with me three years earlier. He even had the same face.

These males were joined by a middle-aged female I named Darki carrying a large male infant I named Derek, and followed by a juvenile daughter, Donika. She climbed much more slowly than a normal juvenile, so I watched her more closely. Her right hand drooped limply and she avoided using it. Had she injured her wrist? I refocused my field glasses and finally saw the shiny, frayed end of a twisted wire protruding from her hand just below the wrist. Another poachers snare. Her hand was crippled but not infected. I wondered if it might become infected as her hand grew, if it grew, but the wire refused to stretch. A horrible thought.

Knuck and Hook, accompanied by a prime male I named Jack, returned to a new *natalensis* nearby. Jack was missing a toe from his left foot. The three males spent five hours eating figs and grooming each other in session after session. Repeatedly Jack grasped Hook's left hand and held it aloft and extended as if to create a hairy A-frame, but actually to groom Hook's armpit. In all the sessions of grooming between Ngogo chimps I had never seen this. They vacated the tree shortly after being joined by five more chimps: an adult female I named Anka, her five-year-old son, Auk, and an adult male with a twisted, emaciated right foot and a withered lower leg whom I named Klubfoot, plus two other males I did not see clearly. Already I was impressed by the number of mutilated apes I was seeing, but there were more to come.

The next morning Hook, Knuck, Jack, plus Darki, Derek, and Donika were back in the *natalensis* along with another mother, Veek (for her V-shaped hairline), who carried a large infant son Vik, and was trailed by a juvenile daughter about Donika's age whom I named Valkerie. With relief, I saw that none of the three was mutilated. This crowd of apes seemed to spur itself to outdo one another in eating, though the infants seemed contented simply playing together.

After an unavoidable week in Kampala I returned to K14 and was happily surprised not only to find two *natalensis* still in fruit, but all of the chimps I had met earlier still in the neighborhood. Soon I met another mother, whose stiff shoulder hairs resembled the puffed shoulders of the Lady Baker dresses still the rage in Fort Portal, despite more than a century having passed since Lady Baker and her husband, Sir Samuel Baker, had been in Uganda. I named her Baker, her infant daughter Beka, and her son Brak. All three were healthy. Knuck joined them with three new males: an old but very muscular male I named King, a prime male who looked like he could crush stones whom I named Kalhoun, plus a thinner older male whose right hand had been amputated below the thumb. I named him Ikarus for the mythical flier who, in his escape from the Labyrinth of Crete, flew too high and lost his feathers to the sun.

By now it was obvious that the Kanyawara Community had more than its share of mutilated chimpanzees. Of the twenty-one I knew well enough to name, five—Hook, Donika, Jack, Klubfoot, and Ikarus—had been maimed. These five, added to the poor juvenile with the snare in her gangrenous foot (who was almost certainly dead by now), plus an adult female missing a hand whom Tom told me he had seen, made seven from a population of less than thirty. Compared to three wounded similarly at Ngogo (Stump, Ita, and Gray [missing only a finger]) from a community totaling fifty-five or sixty, the Kanyawara apes had five times more injuries due probably or certainly to poachers' snares. A quarter of the Kanyawara apes were mutilated, and of these, six were survivors. Because of the likelihood that others had died like the little female, the actual number was probably worse. Kibale was not unique with this problem. Ursula Rahm reported six mutilated victims of snares out of forty-four chimpanzees captured in a forest in eastern Zaire, immediately west of the Mountains of the Moon.

The burgeoning human population surrounding Kibale, especially near Kanyawara, was producing ever more poachers freely taking a serious toll in butchery. Real protection was desperately needed on two fronts: first, *effective* patrols to destroy snare traps and arrest poachers, who then must be *removed* from the forest by prison sentences (not simply fined, which only reduces their profits), and second, a program to educate local people about the irreplaceable goods and services the forest provides them simply by being there, and about the lives of its inhabitants, including the misunderstood chimpanzees. Of course, I still believed Kibale Forest should become a national park, but even then the patrols and education would be necessary.

Meanwhile the Kanyawara chimpanzees were furthering my education. A major misconception concerning chimpanzees is that their sex lives resemble one long orgy. Nothing could be further from the truth for a healthy female. An adult female is rarely receptive to mating except during estrus and the short periods preceding and following it. While this may sound promising from a male perspective, Jane Goodall's (1983, 1986) demographics from Gombe indicate that females nurse their offspring for approximately four years, and during most of that time their estrous cycles are in abeyance and they do not mate. While weaning an infant the cycle resumes, but then stops again after pregnancy. This adds up to each healthy adult female being breedable during only a small number of weeks every five years. And because most communities contain a half dozen to twenty or so fully adult females, the community's males have opportunities to breed only during a few weeks each year. At Ngogo, for instance, only one of the dozen fully adult females—and she only after she lost her infant—came into full estrus, and otherwise I saw only four females that had never borne offspring (Ita, Owl, Quilla, and Zira) in estrus during two years of observations. But now in Kanyawara Anka apparently was weaning Auk; her huge perineum announced estrus. Like a returning comet, her sex life flamed.

One morning I arrived to find the *natalensis* in K14 already providing breakfast to Veek, Vik, and Valkerie, Auk, Hook, and Jack. During the next hour, Knuck, Ikarus, and Kong (a "new" prime male with a gorilla-like saggital crest) arrived independently to bring the feeding aggregation up to ten. When Ikarus arrived, he immediately approached Anka with penis erect. She turned to present her hugely swollen perineum and he mounted her, though for only about two seconds, thus signalling that Anka would be queen for a day.

Almost an hour later Knuck approached Anka with penis erect. She rose to meet him and present her perineum. He mounted her for about five seconds; meanwhile Auk rushed to leap on his mother's back and attempted to push Knuck away. Anka walked forward a couple of yards, and Knuck reached toward Auk, who ducked, screamed, and jumped away. Six minutes later Anka presented to Knuck again and backed toward him, but Auk maneuvered to remain between them and no mating ensued. Then Hook approached to sit in contact with Knuck and near Anka. But soon after he arrived, Kong rushed toward them to grapple with Knuck, who screamed and escaped, and then to displace Hook. Meanwhile Anka fled to sit with Veek, a neutral.

But only ten minutes passed before she climbed back up to sit only within arm's reach of Knuck, who now ignored her. Two minutes later Veek also climbed up to sit next to Knuck, then grasped his left hand in hers to hold it aloft in the A-frame position and groom him for ten minutes. Toward the end of this session, Anka climbed toward another male hidden in the foliage, presented to him, and was mounted by him. Auk was hot on her tail though, and screamed and climbed between the pair, which soon broke apart. For several minutes after this all the chimpanzees huddled in the rain. Eventually Jack and Ikarus abandoned the fig tree.

A half hour later Knuck climbed to plant himself before Veek with his back toward her. It took five minutes before the hint sunk in and she started grooming him—this time for nineteen minutes and once again holding his arm in the A-frame position. But also again, he failed to reciprocate, which perhaps explains her initial reluctance. Meanwhile Anka climbed to present to another hidden male, who managed to mount her for four or five seconds before Auk rushed to pounce on his mother's back and shove the male away.

Soon Anka again approached Knuck, stopped, then backed into his lap. Knuck mounted her and thrust about ten times, and, as Anka screamed, Vik, another infant nearby and Auk's playmate, scrambled to the pair to pull at Knuck's belly, then Auk joined him to discourage the mating. About five minutes later Anka approached another male in dense foliage, presented, and was mounted for seven seconds before Auk managed to climb on her back and try to pry the couple apart. A few minutes later she mated again this way, and after Auk's arrival the pair parted physically but joined in a pant-hooting duet.

Crack, an old female with a juvenile daughter Coke, joined the foragers, and, as the rain finally stopped at about 1 P.M., five social grooming sessions started. Knuck sought out Anka and sat next to her. She began grooming him and grasped his left hand to raise their arms in the apparently obligatory A-frame position. They groomed reciprocally for twenty-nine minutes (Knuck apparently knows on which side his bread is buttered) before Kong approached them. Anka abandoned being groomed to approach Kong, present, and be mounted by him for ten seconds. Again Auk leapt to interrupt, but this time more tardily, having been lulled by his mother's lengthy preoccupation with social grooming. Five minutes later she mated with another hidden male and was harassed again by her son. Afterward this male started to groom Anka, under Auk's careful supervision, but the old muscular King, who recently had arrived, rushed toward the pair, displaced the male, and then mated with Anka for eight seconds before himself settling to groom her. This session lasted only half a minute because, as Veek and her family descended, Anka joined them with Auk in tow. The males remained another twenty minutes, then Kong, Knuck, and Hook descended together. Klubfoot and King climbed down fifteen minutes later.

To summarize Anka's morning, during nearly four hours she mated eleven times with a minimum of four males for an average coupling time of six seconds (Carolyn Tutin [1979] recorded the average mating at Gombe lasted eight seconds). Auk attempted to interrupt nine of those matings, and Vik one (the first short coupling with Ikarus had been too quick to interrupt, and prior to the final one, King had scared Auk away). Whether any of these interferences successfully prevented ejaculation by the male is a difficult question. Male chimps require only a few seconds to ejaculate, and Auk did not actually stop these males from mating; they simply grimaced and put up with his shoving and pulling as if he were a necessary evil. Interference in mating by a female's youngest offspring is the rule not the exception—Jane Goodall (1968) and Anne Pusey (1977) both reported it from Gombe—and it may represent, during the episode when *both* male infants interfered, the earliest age at which a community's males cooperate. But why do they interfere?

If a young juvenile can actually delay the date of his mother's next conception, it gains personally by having her undivided care longer and possibly by nursing longer, especially if it is coercive in its resistance to complete weaning. But by doing so an infant extorts more care than its mother is willing to provide, a situation explored by Robert Trivers in a landmark sociobiological analysis titled "Parent-Offspring Conflict" (see Chapter 12). The source of the conflict, from an evolutionary perspective, is the mother's limited time in which to produce and raise offspring. Nine months of pregnancy plus four years of nursing is what she is prepared to donate to each of her children; more than that might reduce her total lifetime production from four, or five if she is very lucky (Goodall's demographics from Gombe [1983, 1986]), by one infant. (Less than four years of nursing probably would reduce each infant's chances of survival too much to offset the advantage she might gain by producing an extra infant during her lifetime.) But, conversely, the infant being weaned wants more than four years of maternal attention and milk because more may increase its chances of reaching adulthood—hence the insoluble conflict.

Many observations of free-living chimpanzees have reinforced the impression that their mating is entirely promiscuous; this is especially true in provisioning areas where a female's refusal to mate might require her to abandon easy food. Certainly Anka was not keenly discriminatory, but, on the other hand, some young females I observed at Ngogo did refuse to mate with certain males by screaming and fleeing their approach. Although the theory of natural selection generally predicts that a female should chose a superior mate whenever possible because he will donate superior genes to her offspring, among chimpanzees *unique* sociobiological reasons exist for why females *should* be promiscuous.

If male chimpanzees do not emigrate from their natal community, then the males of a community share many more of the same genes than males among most other species of primates. This could be tested by taking blood samples from several dozen males in a population, although the required logistics would be horrible; nevertheless, their genetic similarity is almost guaranteed simply from consistent observations in Tanzania that males do in fact mature to breed only in the community in which they were born. One consequence of this is that the mates a female may choose are more similar to one another genetically and therefore it is less important to discriminate between mates than is the case with other species.

This genetic similarity is also evident in the attitudes of the males toward one another's mating opportunities. Although both King and Kong used their dominance status once each to intimidate other males away from Anka, the general permissiveness, or indifference of the seven breeding males present, was more pervasive than this minor tension. Such indifference makes sense only if the males are closely related such that each male benefits genetically through *inclusive fitness* when one of the other males of his community sires offspring. Of course, these males are not so closely related as to be genetically identical clones so, in the long run, each male prefers to do his own mating. But the genetic competition between them is much lower because of their unusually high proportion of shared genes, and secondarily because tolerance for one another's mating opportunities probably increases their solidarity. On a practical level, because male chimps rarely rely on combat to exclude one another from mating but instead present themselves serially to a female, it is much easier for a female chimpanzee to be promiscuous than to mate with only one male—in short, to reverse the common primate pattern. Despite this cozy arrangement, natural selection will reward most those males who breed the most, hence the strategy of some males to try to take females on exclusive courtship "safaris" and/or to elevate their dominance status (some "alpha" males are able to monopolize some females). Despite safaris and male dominance, however, Jane Goodall (1986) recently reported that during their four days of peak estrus, most females at Gombe mated with most or all of the community's adult males.

But an even more vital reason exists beyond genetic similarity or simply ease for female chimpanzees to be promiscuous. Akiro Suzuki (1971), David Bygott (1972), Kenji Kawanaka (1981), and Toshisada Nishida (1985, with Kawanaka) each independently observed male chimps killing the infants of strange or noncommunity females in Budongo Forest, Gombe, and Mahale. Although the brutal murders of such obviously innocent victims seem atrocious, and rightly so, they made sense sociobiologically. In nearly every case observed, the infants killed had been sired by alien males of another community. And, had they been allowed to live after their mothers immigrated, they would have competed for resources needed by the males' own offspring. An even more pragmatic reason for killing them was to force the sudden resumption of their mothers' estrus cycles which otherwise would have taken years to resume before breeding them would have been possible. The promiscuity of each female *in* the community may be her strategy to gain for her infant the interest and protection of every community male. Of course, none of the males can be certain of paternity, but because all or most of them mated with her, each may suspect he is the father and therefore hedge his bets by treating the promiscuous female's infants as his own progeny. Although this line of logic may seem to demand that male chimpanzees have a solid biological understanding of reproductive processes, it does not at all. Male mammals of most species, even primitive ones, recognize their own mates and avoid injuring their own offspring, but are indifferent or aggressive toward the offspring of other males. The bottom line is this: both male and female chimps in Kibale behaved sexually exactly as if the hypothesis describing their society as based on lethal territorial aggression by kin-related males was correct—and the promiscuity of females would be extremely difficult to explain otherwise.

The following morning I found Ikarus, Knuck, Hook, Kong, Kalhoun, Crack, Coke, Anka, and Auk near the *natalensis* eating *Celtis* leaves. Valkerie joined them an hour ahead of her mother and brother. Knuck managed to mate again with Anka, who was less sexually attractive today, predictably causing Auk to whimper and scramble toward them to interfere.

Kalhoun rushed toward me in a pant-hooting display, then deserted the group to walk southeast. Nearly an hour later two pant-hoots burst from that direction, and, as if a trumpet had sounded retreat, all eleven apes climbed down and headed that way.

Fifteen minutes later I found Knuck hanging back and sitting on the trail. He then joined Ikarus and the other chimpanzees traveling south. I followed, but held back to avoid disturbing them. More than an hour later I found Knuck again on a path. He studied me briefly then knuckle-walked southeast again, having gone only about four hundred yards in an hour. Twenty-five minutes later and another several hundred yards south I found Knuck a third time, now in K30. Most of the party now had stopped to rest quietly in the dense understory near ground level. Knuck spotted me and moved closer, then closer again. By now I was convinced that he was the same young subadult of three years earlier.

After twenty minutes Knuck rushed toward me, pant-grunted, then halfheartedly waved a branch at me, as if experimentally. Probably disappointed at my response of merely scratching and glancing away, he climbed to a low branch nearby to sit dangling his legs and observe me. Swishes in the foliage revealed that most of the apes were drifting further south. Screams a few hundred yards in that direction were chorused about a minute apart. Knuck emitted a small "yip" after the second one, then, a few minutes later, dropped to the earth to follow the others.

Nearly an hour later and a mile further, now at the southern boundary of the K30 trail-grid system, I found seven apes of the original party in K14: Crack and Coke, Kong, Knuck, Anka and Auk, and Kalhoun, plus five others who had joined them after the initial move: Baker, Beka and Brak, King, and Jack. At this same moment, Joe Skorupa was searching for red colobus in K14 and saw Hook and Klubfoot close to where the apes had gathered during the morning. Klubfoot's withered leg and Hook's crippled hand made travel difficult; both had avoided the three-hour, mile-plus trek, but at the expense of solidarity, and of Anka. I wondered if they knew in advance that they were avoiding this walk by splitting from the main party and, if so, *how* did they know?

Most of the apes were scattered in the lush understory and merely sitting as if killing time in anticipation of some event scheduled in the near future. Soon Kalhoun used his canines to peel long strips of bark off a *Markhamia* tree, then ate the inner bark with evident relish. Inspired, Knuck followed suit, gnawing out similar strips with his temporary canines. Other chimps gathered into small grooming clusters.

A buttress-drumming and brief pant-hooting from a couple of hundred yards west elicited an immediate chorus of pant-hoots from Jack, Knuck, and two other males. The females just groomed. Almost twenty minutes later, Anka pant-grunted a husky greeting as Ikarus emerged from the greenery in the north-west. With his mutilated, fingerless hand, he was probably slower than the others. Had that recent buttress-drumming and pant-hooting been his way of asking for help in zeroing in on the group? As if his arrival was their cue, all thirteen apes plodded beyond the trail-grid system toward Ngogo. About ten minutes later I heard a chorus of pant-hoots a quarter mile into this soggy terra incognita. I wondered where they were going.

Two hours after dawn the following morning the females, Anka and Baker, who had disappeared south, had returned north with their families to join Darki and her family in the fruiting *natalensis* in K14. This necessitated at least a mile and a half walk for breakfast fruit. I realized now that Darki's family, with poor Donika crippled by a wire snare, were the only other individuals besides Klubfoot and Hook who had split from the traveling party yesterday. It now dawned on me that all three defectors were cripples; this was my first inkling that such wounds could have such serious social repercussions. The depredations of poachers were not simply killing and mutilating the apes, but altering their social interactions.

Oblivious of their elders' problems, both infants, Beka and Derek, joined with Auk for a half hour of chasing, wrestling, tickling, and mock-biting one another in the crown of the fig tree. Their husky panting laughs sounded like desperate gasps for air.

As the sun intensified, all three families descended to linger in the shade of the forest floor. Soon a deafening chorus of pant-hoots erupted from the base of the fig tree, then Klubfoot, old King, and young Knuck climbed into view as if scheduled re-placements for the families who had just descended. Whether the latter two males had returned from south of K30 for figs or a chance to mate again with Anka is anyone's guess, but they foraged only half an hour before vanishing into the understory. By the early rain of afternoon both figs and chimpanzees became scarce, and I had to search the region again for alternate foods.

I found King and Hook, who for the next three hours before

dusk remained only a few yards from one another while they lolled about, self-groomed, and ate hundreds of young *Celtis* leaves. After sundown the pair descended to plod through the understory. Ten minutes later they chorused a pant-hoot and were answered by a single male, roughly a hundred yards north. These calls signaled the end of an exceptionally good week with the Kanyawara Community. My searches the next day failed, and I had to return to Ngogo.

My weeks with the apes of Kanyawara had been rewarding beyond my greatest hopes. Other than having revealed patterns of mating, tighter social solidarity during travel and other times, plus a shocking number of mutilations from snares and their repercussions on social life—and, for that matter, on survival—the Kanyawara chimps had convinced me that they definitely differed from their cousins only less than ten miles south at Ngogo.

One black and white difference was the Kanyawara chimps' habit of grasping a grooming partner's hand in the A-frame position during 38 percent of their sessions of social grooming. Although this did expose the underarm of the groomee, sometimes that area was not even groomed. It may have added stability to the pair in some circumstances, but at other times it seemed to make them more precarious. More than anything else, the A-frame hand clasp seemed a custom or tradition, perhaps transmitted over several generations. Then again, it may have been only a fad. But I never saw the apes of Ngogo do it during 200 sessions of social grooming.

William McGrew and Carolyn Tutin, veteran observers of Gombe chimpanzees, visited the Mahale chimps and noticed the identical difference in behavior: the Gombe chimps never clasped hands in the A-frame position, those at Mahale did frequently. McGrew and Tutin assumed that isolation (at least 30 miles of unsuitable habitat between the closest portions of the apes geographic ranges) over many generations contributed to this anomaly, which they proposed was a valid but rare example of a naturally occurring *cultural* difference among nonhumans.

No such vast hiatus, however, separates the chimps of Ngogo from those at Kanyawara; a determined ape could walk the few miles between the two areas in one day (I have hiked it in three hours), though the route might include a mile or so of colonizing vegetation. This difference prompted me to wonder anew about the degree of intercourse between the two communities. Behavior as simple as the hand clasp would show up easily if a female emigrating from Kanyawara settled in Ngogo, unless the apes there resisted it through being unltraconservative. The Ngogo chimps' tendencies to travel seasonally south from Ngogo, away from Kanyawara, plus the large area between Ngogo and Kanyawara, both left room for a potential *third* community of apes between them. Unfortunately I was never able to investigate this possibility.

The habitats of the two known communities differed: Ngogo had dozens of giant *mucuso* trees and thousands of *Pterygotas*, sources of the two top ranking foods of the apes; Kanyawara had only one *mucuso* and very few *Pterygota* trees. Ngogo offered a wider expanse of forest, while Kanyawara was a narrow peninsula of forest surrounded except in the south by disturbed, degraded, or destroyed habitats including partially felled forest like K14, arboricided forest north in K15, artificially maintained grasslands, plantations of exotic monocultures such as Caribbean pine, cypress, and Australian eucalyptus, plus ever-increasing *shambas*, people, and rampant poaching. One immediately obvious effect was the five times more frequent mutilations among the Kanyawaran apes caused by poachers' snares. This difference was not intrinsic like the hand clasp, however; it was forced on them by human culture.

Besides the hand clasp, the Kanyawara chimps seemed much more cohesive in their traveling parties. Perhaps as a consequence, dominance interactions were five times more frequent than at Ngogo! But their antidote, reassurance, was much more common as well. To what degree are these differences simply responses to different environments? Or, to what degree are they *socially transmitted behavior patterns*, in other words, *culture*?

First, no doubt whatsoever exists that the behavior of chimpanzees is modifiable. In addition to the examples above, many observations of the apes in different habitats demonstrate this. In fact, in my opinion, the many differences in food types and processing between the Gombe and Mahale chimps discussed in detail in Chapter 4 provide several examples of cultural differences. In this same vein Yukio Takahata, Toshikazu Hasegawa, and Toshisada Nishida even found substantial differences in the

rates of predation, prey types, and sexual roles of the chimpanzees who hunted in Mahale versus Gombe which probably
also are cultural.

Although they may seem extreme, the ultimate examples of
how easily a chimpanzee can adapt to arbitrary routines come
from captivity. Cheeta, captured as a tiny infant in Liberia in
1936 to star in thirteen Tarzan movies and several other films, is
described by his lifelong trainer, Tony Gentry, as "just like a
member of the family. He comes in the house and has his coffee
with me. He gets his bacon and eggs." Now fifty years old and
semiretired, Cheeta spends most of his time watching television,
drinking generic beers, smoking cigars, and "aiming the smoke at
the ceiling."

The big question is not so much malleability; it is adaptive
value and cultural transmission. Clearly behaviors associated with
foraging, hunting, grooming, dominance, and other aspects of
chimpanzee life vary from community to community, but how
adaptive are they? Do they somehow allow the community's
members to increase their reproductive success? And, if so, are
these differences transmitted socially from generation to generation? Do they constitute *culture*? The current state of the art
does not allow us to flatly assert that such differences increase
reproductive success, though in some cases logic suggests they
do. The A-frame hand clasp may be only a fad, but some differences in food processing, such as pounding hard-shelled foods
with rocks, or hunting cooperatively, may be highly adaptive
innovations. Jane Goodall (1986) and William McGrew (1977)
reported that it was critically necessary for immature chimpanzees at Gombe to watch their mothers and other elders to
learn various potentially *cultural* techniques of food processing,
tool construction and use, and even some social behaviors. In
addition, immature chimpanzees were the most likely to ignore
their elders' conformity by experimenting with new foods or
other items. It is an excellent bet that, if a juvenile male from
Kibale were sent as an exchange student with Cheeta in Los
Angeles for a year or two and then returned to Kibale with
continuous supplies of generic beer, future generations of Kibale
chimpanzees would be drinking beer as long as supplies lasted.
The propagations of new behaviors, neutral or better, is a virtual
certainty between chimpanzee generations, and, of course, so is

culture. Assessment of their adaptive value, however, remains a serious challenge.

I am convinced the Kanyawara chimpanzees' greater social cohesion and more intensely active life-style is a direct reaction to the increased degradation of their habitat and their concomitantly greater need to exploit those resources remaining as efficiently as possible. A greater reliance by each individual on other apes for their combined knowledge of where and when the best food resources are to be found could easily explain their greater cohesiveness. If true, their differences in life-style from the apes of Ngogo are adaptive. And, if they survive, these differences will become *cultural*.

I had hoped to develop this assessment further beyond speculation, but, five months after my return to Uganda, the new regime of Ronald Reagan excised my project from the budget of the National Institutes of Health.

One major lesson in all this is that to generalize about great ape behavior observed in only one group or community is somewhat risky. This lesson already was clear to me in December of 1977; it not only broadened my outlook, but intensified my interest in the apes of Ngogo, who despite being more difficult to observe, provided an excellent model for an undisturbed, steady-state community of wild chimpanzees. Unfortunately, this steady state was seriously threatened.

— 11 —

Lightning War of the American *Chui* Battalion

In January the rains fizzled to nothing. The sky faded to dirty white, then brown as farmers on the fringe of the forest and beyond set fire to the stubble of their harvested crops. The load of ash in the air was increased by pyromaniacal poachers moving onto the grasslands within Kibale Forest. A green sward arises from the ashy soil two weeks after the grass is put to the torch, even without rain. As predictably as the grass, these islands of forage become dotted with grazers. Again predictably, hunters lay in wait, their nets set up for hundreds of yards, their illegal firearms loaded with black market ammunition, their iron-headed spears honed keen. The dry season and fires transformed the grassland into a cornucopia for poachers.

This January was my fourteenth month at Ngogo, and so far the Kibale Forest game guards had not arrested a single poacher of animals. Even though the game guards were badly outnumbered, it seemed suspicious and I sometimes pondered their integrity. But because I was an alien among the Batoro, it was impossible for me to come up with straight facts that might vindicate or impugn the game guards' professionalism. I felt an uneasy guilt over my doubts about these men. They risked their lives against unfavorable odds to protect animals they craved as food and unexploited land that would make them wealthy if they owned it. It was a lot to ask of anyone, let alone of someone whose wage did not cover the basic necessities.

But this January the antipoaching stalemate finally broke. The Uganda Game Department assigned another game guard, Makuru, and his assistant, Nyakana, to the central block of Kibale Forest. Tom and I were hopeful that things would change. They did, but the ensuing changes were complicated by the fact that Makuru had grown up in Kanyawara, one of the areas in which he was to patrol, and had been raised in a family for whom poaching was an adjunct to farming.

I pieced together the following episodes of the January "lightning war" from my individual debriefings of the guards. While their veracity is not certain, they held up under cross examination and testimonies from other witnesses. Their blitzkrieg illustrates much about antipoaching work that normally does not see the light of day.

On January 18, 1978, Mubu and Mukasa were joined by Makuru (Otim was on leave) and Makuru's assistant, Nyakana, and set off through the forest along a maze of paths stamped into the earth by a new but vanishing breed of miniature tusked elephants created by poachers' overkill. Near Rwetera, a tiny village and tea plantation seven miles from both Ngogo and Kanyawara and situated on the western fringe of the forest, they found improved paths hacked through the foliage and superimposed on the elephant paths. At noon, fresh spoor led them to a pair of red duiker carcasses hanging in a tree for later butchering. The four knew the hunters would return to claim them, so they encircled the carcasses and concealed themselves.

Six hours passed. Each man became lost in his thoughts, then relost in the same thoughts until new thoughts were harder to come by. But with fatalistic patience typical of Africa, they remained in hiding.

A faint chorus cut through the still air, then grew louder. Those waiting slightly shifted their positions.

"I was lucky to spear it," one of the three approaching said as a prelude. "I'll take one leg."

"I'll take the head, lung, skin, ribs, and another leg," a second man added. "It was my net."

The third man, carrying a third antelope, began to discuss which portions of it would go to the headman of the hunt, but then froze in his tracks. While bodily hidden, Mubu had crouched with his rifle vertical. The third hunter was now staring at the barrel of Mubu's rifle projecting above a bush. All discussion ceased.

Abruptly Mubu stood to warn them not to attempt to flee. The trio dropped the spoils of the hunt, plus three spears and three nets, then bolted in three different directions. All four game guards gave chase but almost immediately heard the voices of more men approaching, so they quickly turned back to conceal themselves again around the hanging duikers.

The trio now fleeing had been only a vanguard of a group of eighteen poachers out on a cooperative hunt. Their jests and shouting grew louder in the ears of the waiting guards. When the hunters were thirty-five yards from their rendezvous, Mukasa rose to his feet, fired a round in the air, and shouted a warning to stop or he would shoot.

Fifteen nets, along with sundry other appurtenances of the poaching trade, thudded softly to the forest floor. Thirteen hunters fled into the forest as fast as their legs could carry them. Two men dropped to the earth to avoid being shot. The game guards apprehended these, then Makuru covered them with his rifle while the other guards gave chase.

After a half hour of unsuccessful pursuit, they returned burdened under fifteen nets and other impedimentia they had gathered on the way back. They tied the two captives together, then commanded them to gather an immense heap of firewood for the long equatorial night.

As the sun vanished beyond the Mountains of the Moon, they lit a fire using Double Happiness matches, a product of Mainland China ubiquitous in Uganda. The six men sat close to the fire and gorged themselves on the timid red antelope that had been transformed to *mnyama* (meat), then to legal evidence, and once again, through necessity, back to *mnyama*. The two poachers received the best cuts of evidence because they had done the hunting. Tradition runs deep.

The game guards gradually fed seventeen of the eighteen nets into the fire because they had no way to transport them to Fort Portal. They retained the final net as evidence. Then they partially covered themselves with animal skins and hunched close to the fire to wait out the night. In the morning they would start the trek to Kanyawara. Public transportation from Rwetera, had such been available, would have been preferable, but the Game Department did not reimburse its guards for travel expenses, even to bring suspects to the Fort Portal Police Station. They had to walk.

At dawn Mukasa walked back along the path of the poachers' retreat to search for any items missed during their hasty sweep of the previous evening. Less than fifty yards from the fire he discovered a stranger curled in slumber at the base of a tree. His spear was tucked against him. The man had become separated from his companions and had been fearful of traveling through the forest at night. Having backtracked to the fire he felt safer, but he had been unsure who the proprietors of the fire were. He suspected they were game guards but was too prudent to go closer and find out, and too frightened of the forest to flee. He compromised by staying on the fringe of the fire's protection. His plan was to leave at the first hint of dawn to avoid contact with those tending the fire.

Mukasa quietly slid the spear away from the recumbent hunter. Then he slapped him three times on the ear. (Mukasa grinned as he retold this.) Suddenly aware that he had overslept, the man jerked his eyes open to have a .375 caliber rifle offer him his morning greeting. Muskasa, tickled by the man's predicament, yelled to his comrades that he had just captured another poacher. They shouted back in disbelief, "How can you find one so close?"

Mukasa led the befuddled hunter, who probably was hoping that all of this was simply a bad dream, to the fire where he was tied to his colleagues and fed some evidence. Twenty miles by road, and nearly eight hours later, they walked into the Kanyawara Forestry Station where Tom Struhsaker drove them to Fort Portal. Due to an unusual judicial distaste for poaching, in less than a week each of the three poachers was convicted and sentenced to a year in prison.

Six days after their march from Rwetera, the game guards patrolled together again, this time in a peripheral region of the forest northeast of Kanyawara and directly east of Rwetera.

Loosely referred to as the Kiko Side, it was a jungle of second growth forest competing for space made vacant by selective felling of the east boundary and by "refinement" projects of the Uganda Protectorate. "Refinement" is a bureaucratic euphemism for mass poisoning of tree species which do not produce commercial timber, including fig trees—a messy and ecologically absurd practice. The region was penetrable by man only because of elephant paths, poachers' paths, and a single access road that was being recut from near oblivion by Okwilo, one of Tom's cutters.

Three of the guards forged ahead, leaving Mubu to guard Okwilo, who had seen many recent signs of poachers and knew he was in danger because he represented, albeit in only a small way, law and order. Makuru, Mukasa, and Nyakana came upon a path bearing fresh slash marks made by a *panga*. They followed this and soon found "foot marks." After a mile of stealthy tracking, the prints led to the carcass of a freshly killed red duiker hanging in a tree. Déjà vu.

The three concealed themselves around the dangling prey and waited. Ironically, the three poachers who had cached the antelope were dismantling their trap line to move it to a new location. The recent cutting by Okwilo had convinced them the region was becoming too risky. They knew the game guards would confiscate any snares they discovered, requiring them to spend their shillings on new wire. Although unlikely, the possibility of being caught was also real. After gathering thirty-two snares, the hunters returned to retrieve the duiker they had found in one of their snares earlier that day.

A snare trap is more sophisticated than a bent sapling with a wire loop at the end. A shallow trench is excavated and sometimes bordered with two guide rails of small logs so that the foot of the prey will sink cleanly into it, through the loop of the snare, and onto a trigger. The trigger is attached to a wire leading to a toggle holding tension on the bent pole with the main snare wire attached to it. Pressure on the trigger pulls the toggle loose, releasing the tension bent pole, which snaps upright and clinches the snare wire within the pit by pulling from one side under a guide rail (if there is one). When an animal steps into the trench, which has been camouflaged with leaf litter, it is rarely able to lift its foot back out again. Of all the local industries I observed in

Toro District, these traps seemed one of the most inventive. (Months later, in this same area, I stepped into one of these traps. But because my boot was too large for the loop, I was not snared.) My purpose in describing these snares is to illustrate the amount of investment the poachers were giving up by dismantling their entire trap line.

The cautious trio jammed the thirty-two wires among the roasted sweet potatoes in duiker skin bags strapped to their hips, then weaved a path back to their cache. They arrived quietly about an hour after the game guards. At first the hunters seemed nervous, as if suspicious of some change in the place. But this time no rifle barrel projected above the foliage. After a brief pause they unstrapped their waist packs, retrieved the duiker, and started to butcher it.

All three game guards stood and shouted, "*Simamisha au tuta-piga* (stop or we will shoot)!" The effect was galvanic. The hunters dropped everything except their spears and bolted for the jungle.

Makuru fired a round as he rushed after them. Fearing a bullet in the back, one hunter dropped prone against the earth. Makuru closed on him and yanked the man's *panga*, his sole weapon, from his grasp. Then he held him at gun point.

Mukasa and Nyakana raced after the other two, who soon separated where the trail forked. One followed each of them. Nyakana's man stopped suddenly and, with a lightning twist, launched his spear at him. Nyakana leapt aside and shouted. Because he was not a game guard but only an assistant, he was not authorized to carry a firearm. He carried a spear instead, the same type of spear favored by local poachers, except that Nyakana's had the heaviest, sharpest blade I had ever seen in Toro. In fact, because during the harsh regime of Idi Amin the Game Department ceased issuing uniforms to its men, Nya-kana'a tattered bush clothing made him and the game guards indistinguishable from the man he was chasing.

That poacher's spear whizzing past him dimmed Nyakana's ardor for the chase. Perhaps the wage he received, the minimum allowable in Uganda, worth about twelve U.S. dollars each month in real buying power, seemed inadequate recompense for being speared. Perhaps no wage would be worth that risk without the advantage of superior armament to provide the winning edge in the hand-to-hand combat so likely during antipoaching patrols in thick bush. Rather than press the chase, Nyakana turned and rushed back along the trail to turn off on the path down which Mukasa had just vanished in hot pursuit of the third poacher.

Mukasa saw that he would not be able to catch up to the man rushing ahead of him. His best chance of stopping him seemed to lie in stunning him by firing a bullet over his head. Mukasa slowed down to lever a bullet into the chamber. The bolt jammed. Feverishly Mukasa tried to lever a second round into the chamber, but it would not feed, and the bolt jammed again.

As Mukasa loped along the trail with his attention divided between trying to keep the man ahead of him in sight, trying to avoid becoming entangled in the thick vegetation, and trying to convince his uncooperative rifle to accept a bullet, his quarry glanced back over his shoulder at him. When the hunter saw Mukasa struggling with his rifle, he stopped and turned to face Mukasa and make a stand.

Mukasa glanced up from his rifle to see the poacher facing him. Simultaneously his rifle was torn from his hands. The bolt had caught in a hanging liana, and his momentum had carried him along the path toward the waiting hunter. The rifle dropped to the ground behind him. The hunter lifted his spear and poised to launch it at Mukasa, who now was in point blank range.

At a loss for any plan that might take him safely from the bite of that spear, Mukasa continued to run at full tilt into the man about to throw it.

For reasons unknown, the spearman hesitated. It is no small thing to kill a man, no matter how many lesser creatures one has killed during hunts. A man is different. Perhaps the hunter was stunned by the enormity of what he was about to do. But that instant of hesitation was long enough for Mukasa to close on him.

Over they went in a frantic tumble of groping limbs. Mukasa wrestled his way astride the man to gain a grip on his throat. At such close quarters the spear was useless, but the hunter had retained his *panga* and now, with Mukasa trying to strangle him, he tried to use it. But as he reached back for a slashing stroke that might have killed Mukasa, he grabbed the man's wrist. Wresting away the ubiquitous knife of the jungle, Mukasa desperately hacked it across the hunter's head. Blood flowed from a slice running across forehead and scalp.

Nyakana appeared running along the trail. He paused to pick up Mukasa's reluctant rifle. Mukasa yelled for the rope Nyakana carried to truss up captured poachers. Once the wounded prisoner's hands were tied, they returned to the duiker cache.

Mukasa's panicked struggle caused him to revise his opinion of the position of game guard. Before his brush with death he considered it a good job. Beyond his meager salary of sixteen U.S. dollars per month, the spoils of the chase—food, blankets, cooking pots, and other useful items—were his when he arrested poachers or caused them to drop their possessions to flee more swiftly. But he had just felt the icy breath of death and did not like it. Later he told me he did not consider his job to be a good one.

They tied the wounded poacher to his comrade, then marched them to Kanyawara toting the snares and everything else of use. In Fort Portal the police told Mukasa that no questions would have been asked had he killed the poacher during their scuffle. After all, the man had tried to kill him.

Three days later I plodded up the hill at dusk to Ngogo Camp after a day in the forest. I was contemplating an adult male chimpanzee and his juvenile male companion, whom I had been watching for nearly three hours deep inside the forest. Male chimps, like these two, often showed a close companionship that never ceased to intrigue me. I wondered again what degree of territorial defense, if any, was spurring these all-male parties I kept seeing. My musings were scattered as Mubu rushed to greet me. Behind him the big hill was crowned by a wavering curtain of smoke.

Speaking quickly, as if standing on a hot surface and eager to step off, Mubu explained that he, Mukasa, Makuru, and Nyakana had captured three more poachers. The game guards wanted to transfer them immediately to the Fort Portal Police Station and wanted me to drive the forest track. I countered by pointing out that nightfall was upon us and driving the forest track in the dark was too dangerous. I loaned him my watch and explained that the four of them could set a rotating watch to guard the prisoners during the night.

Mubu marched the captives, trussed together, into the forest to gather an enormous load of dead wood for the night fire. After directing this expedition, he and then the others described to me the events leading to the arrest.

An hour after dawn that morning, the four had patrolled south from camp along the grassland track to scan for poachers from the hilltops. About four miles away they climbed the large hill called Miko. From there they spotted a tendril of smoke rising from the big hill from which I had first viewed Ngogo Camp with Simon. The fire had been kindled literally in their tracks. They hurried back to it.

By noon they were searching the big hill for a sign of those who had set the fire. They circled to the base of the hill where the fire appeared to have started. Makuru soon found broken stems of grass and footmarks leading east.

Roughly a mile from the fire the tracks led the guards to an exposed game pit, a V-shaped hole about twenty feet long by only a yard wide and more than two yards deep. The normal camouflage of branches and leaves was scattered nearby, along with offal from a butchered bushbuck. Hooved animals who step on the flimsy cover of these pits slide to the nadir of the "V" where all four hooves jam in a straight line. Constricted thus, they cannot maintain their balance to leap up the steep side. This particular bushbuck evidently had died of thirst long before the poachers found it; the hair was already slipping from the hide. Normally the hunters would have recamouflaged the pit after hauling out their catch, but these had been in a hurry to visit several more pits. At this point their pursuers were only an hour behind them.

Continuing to follow fresh tracks, the four dogged a sinuous path through prickly *Acanthus* scrub. The trail continued for more than a mile through forest to finally emerge onto grassland. Soon the foot prints veered back into the forest again and led along a route parallel to the interface of the forest and grassland. Reckoning that they must be getting close, each man became more cautious, speaking rarely and only in whispers.

WHACK! The sharp ring of a *panga* hacking wood came from a short distance ahead. By now the four felt like veterans in the man-hunt trade; they quickly deployed into crescent formation and advanced toward the wood chopper.

The poachers had not anticipated being followed and had stopped near the edge of the forest to kindle a fire and set some of the semiputrid venison over the flames.

The four guards moved in so close that the voices of the hunters were not only audible, but intelligible. But the foliage was so dense that none of them was visible. The guards were unsure as to how many men they were about to face and what weapons they had. By relying on precedent, Mubu broke the tension by firing his rifle in the air and shouting a warning not to run.

Four poachers bolted in four different directions; three scrawny dogs scurried after them. Makuru saw one man run and sped after him. The fugitive seemed as fleet as bushbuck. He came to a small clearing and leapt across it. Makuru crashed along behind, rushed into the clearing, and, as he ran, the earth beneath him disappeared.

He felt like a fool as he picked himself and his rifle up from the litter in the bottom of another V-shaped game pit and scrambled up the side to peer over the edge at the fleeing poacher. Makuru's foot slipped and he slid back into the pit again. In desperation he fired off a round from the bottom. He then tried climbing out again and saw that the poacher had dropped prone, fearing a shot in the back. But now he inched his head up and saw Makuru again struggling to escape the pit. Seeing his chance, the hunter rose to his feet and sprinted toward the forest. Makuru was beginning to suspect that he might never escape the pit, so he fired a second shot. The poacher dropped prone again. Makuru clambered out of the pit, rushed over, and arrested him.

Meanwhile Mubu gave chase to a man who, spear in hand, ran across the grassland. He glanced back to spot Mubu, then veered to the cover of the forest. Mubu ran harder, his tire-tread sandals pounding against the drying grass, but his quarry entered the tangled vegetation of the forest–grassland ecotone well ahead of him. Seconds later Mubu entered the forest and fired into the air. The fugitive dived to the ground and remained prone, still gripping his spear.

Mubu hurried to the man, grasped his arm, and commanded him to release his spear. The man struggled against Mubu, who is six feet tall, much larger than the average Mutoro, so Mubu slapped him in the face. This blow only intensified the man's refusal to drop his spear. Mubu swung the butt of his rifle down hard on the hunter's head once, then again so hard that both the scalp of his prisoner and the stock of his beloved Winchester .375 were split open. Mubu wrested the spear from the screaming man. Nyakana heard the screams from a quarter mile away and jogged over with his rope. They trussed the man, then marched him back to the cooking fire.

As they neared the fire they heard another shot. They left the prisoners to be guarded by Makuru and rushed in its direction. A few minutes later they met Mukasa leading a third hunter back to the fire. He was carrying the man's spear. The arrest had been somewhat standard: a long chase and a shot fired to scare the runner into hitting the earth. Mukasa was becoming an old hand at this sort of thing.

Back at the fire Mubu shot the female dog, whom he said was the hunt leader; the others fled again. By interrogating their captives they learned their home was Mbale Village on the Mwengi Side, along the eastern fringe of Kibale Forest. (A month later Nyakana happened to pass through this village; upon being recognized by one of the current prisoners, he was beaten to within an inch of his life by a gang of the man's cronies.) The four poachers had left their village the previous morning, a Saturday and prime time for poaching, and had spent the night in the forest. In addition to parts of the bushbuck, they were carrying the meat of two red duikers, one of which had been pregnant, that they had also found in game pits.

They led the captives back to Ngogo roped together at the waist "like slaves," and carrying over twenty pounds of meat plus other poaching paraphernalia, including a rope holding a duiker horn container, three carved wooden holders, and a bag containing medicine to ensure the success of the hunt.

After explaining all of this to me in Swahili and English, Mubu, Mukasa, and Mukuru retired to the bonfire to consume evidence. Meanwhile I stripped Mubu's Winchester, cleaned the stock, and repaired the crack with epoxy glue. The prisoners spent the night roped together in the kitchen hut. After nearly an entire night of the game guards talking and laughing around the fire, dawn finally tinted the smoke-filled sky a dull crimson. We loaded the

Land Rover with all concerned and I aimed it up the charred hill. The vehicle bounced over boulders half buried in the soil and swayed drunkenly. Almost immediately the left rear tire went flat.

Changing it should not have been a problem, but Land Rover lug wrenches are short and made of soft steel. Inconveniently, the toughest man in the Ministry of Works had tightened the lugs on this particular wheel using a wide star wrench. The Land Rover wrench now bent like rubber when I tried to remove them. Eventually the wheel was held by one last nut that would not budge. How ironic, I mused to myself (knowing no one there would ever understand the pun), justice foiled by a little iron nut.

I doused it with oil, then I used a screwdriver as a cold chisel by pounding it at an angle tangential to the nut's rotating axis. The tool slowly disintegrated. The three captives squatted on one of the wheel ruts on the track and gazed into the distance as if in a trance. The game guards offered me encouragement in Swahili and English and speculated with one another over the cleverness of my approach. Mukasa took a turn hammering the screwdriver. I took over again. We swapped back and forth for an hour and finally destroyed the screwdriver—just as the lug finally broke loose.

"Ah, very scientific!" approbated Mubu. He considered me a *fundi* (expert), and called me a genius, because I had been able to repair his rifle. The Game Department did not teach their guards to repair their weapons, or even to field strip them to clean them. Their weapons training consisted only of firing three shots.

As we bounced along again on the fresh ash Mubu explained to me the name the game guards had given themselves. "We are the American *Chui* Battalion. You understand *chui*?"

"Leopard," I answered.

"Ah, yes. You know Swahili very well" (which I really did not).

"Why *chui*?"

"Ah, because, you know, *chui* cannot give up. I mean he cannot be defeated—he is too strong."

"Good name. Why American?"

"Ah, because we are paid by Tom Struhsaker, who gets the money from America."

"Oh." I wondered about the repercussions of their self-chosen unit designation with the Game Department, whose pride might be adversely affected by the support unit of guards attached to Kibale considering itself something apart from the Uganda Game Department.

"You know, if Tom goes from this forest, it will be finished. No one cares about this place. Those poachers will come and finish the animals, and men with saws will cut the trees."

"Sometimes I think you may be right."

Two hours later we rolled off the smooth pavement of Fort Portal's roadway into the parking lot of the police station. The transition from Ngogo to the slick new Africa of Fort Portal always seemed surrealistic to me.

"Mubu! You are bringing them daily!" the starched and slightly overfed booking sergent behind the counter congratulated Mubu in English. Then, in a much quieter voice, in Swahili, asked. "Is there much evidence?" The game guards of course had nibbled the evidence down to a quantity adequate for conviction but not for consumption.

"Very little," replied Mubu.

Mukasa slid the evidence across the counter, including spears, meat, and everything else. The sergeant fingered them, then picked up the rope with the duiker horn and wooden containers of magic to ensure a successful hunt. He turned to the prisoners and then to the guards and asked, "What is this?" as if to emphasize that he, for one, was light-years beyond believing in magic talismans.

Then he instructed the suspects to stand behind the counter. Another of the policemen commanded, "*Cheza . . . Cheza* [dance]!" All three reluctantly began to sway slightly to some internal rhythm. Dressed in scant rags, torn and dusty from the bush, one man wearing a basket for a hat, another bleeding from a scalp wound, all three shrinking within themselves and almost visibly attempting to disappear, they seemed wild creatures torn from their place in nature and trapped in a hostile, unnatural environment.

"*Imba* [sing]!" the policeman commanded. When they simply looked at him he prodded them.

They murmured a sonorous and monotonous chant to accompany the shuffling of their dry, dusty feet against the concrete.

The sergeant then commanded them to clap their hands, and they did. All three were singing their mournful chant, swaying, and clapping their hands, while their hollow gazes probed a distance we could not see.

I felt a surge of empathy for them. In their tattered rags, equipped with spears, *pangas*, grass baskets, burlap sacks strapped toga-like across the chest, a rusty old tobacco tin to keep their Double Happiness matches dry, and a bit of rope for magic talismans, they seemed elemental hunters of the bush, a vanishing species in their own right. But they were nothing of the kind.

They were weekend poachers equipped with their most worn and expendable trappings. They had *shambas* and wives back in Mbale. Their incursions into Kibale Forest were not a means of staving off starvation, but of procuring extra protein and perhaps extra cash from the sale of meat back home. Their exploitation of Uganda's vanishing wildlife was a classic example of the tragedy of the commons (the "commons" being a British term for grazing land used in common by numerous herdsmen). This process, if unchecked, will lead to the total obliteration of the resources being exploited.

The men of most communities once traditionally hunted freely in the wilder areas surrounding their agricultural villages. When their populations were small, so was their impact on wild populations. But now, with Uganda's human population burgeoning in the wake of Western medicine and most populations of wild game wiped out through overkill, the only remaining populations worth hunting are in the few remaining reserves protected by law, including only three national parks. The villagers living outside these lands look upon them as hunting grounds for the common use of all willing to brave the law by poaching. Every group of hunters is faced with the same simple choice: kill another animal and gain *all* benefits from the kill, or abstain from the hunt to allow the animal population to hold its own, which also benefits hunters but by a factor greatly diluted by that benefit being *shared* in common by all other groups of hunters. Naturally they choose to hunt; it gains them the most in the short run and it is the only option producing results of which they are sure. Sociobiologically, because wild game is the common property of everyone, it is in each hunter's immediate, unenlightened, "best" self-interest to kill what he can before other hunters do. Harvesting of this type, called scramble competition, is the antithesis of the careful culling of individuals from domestic herds which allows the herd to flourish despite the deaths of chosen members. And it occurs because wildlife populations have no individual owner with a vested interest in ensuring that the reproductive base of the population remains intact. The Game Department *is* (nominally) this single owner with vested interest, but its anemic budget and flagging esprit de corps essentially has given license for the sociobiology of hunters to create this tragedy of the commons—a tragedy indeed once species have vanished forever into the black hole of extinction.

These poachers in the police station were the same men who joined in organized bands who set up nets end to end, then drove frantic forest creatures into them to be met by spearmen. They allied themselves with rifle-hunters who already had killed the last hippo, lion, and probably leopard, and had nearly extirpated the waterback, buffalo, and elephant. They set up indiscriminate snares condemning thousands of animals to cruel, prolonged deaths—or mutilation if they were lucky enough to escape.

To resolve that these men *must* be stopped I had only to remember the many maimed apes in Kibale, particularly the pathetic little juvenile female chimp dragging that hunk of gangrenous flesh with a snare wire embedded in it that used to be her foot.

A few days after their arrest, the three were released without even the nominal requirement of paying bail. A few weeks later, through some influence with the prosecutor, they were charged only with the minor crime of carrying spears in a forest reserve. The court found them guilty and fined them one hundred shillings, equal to about three U.S. dollars. Their fine failed to cover even the cost of the gasoline used to transport them to the police station. Thus ended the lightning war of the American *Chui* Battalion.

During February Mubu went on leave. Lysa drove him into Fort Portal where he caught transportation to his home nearly a hundred miles south. She offered to take him to the police station to turn in his rifle before going on leave, in order to comply with the Game Department regulation prohibiting game guards from taking their weapons on leave. Mubu insisted, however, that he would turn it in later, but he departed Fort Portal without visiting the police station.

Upon return from his leave, Tom asked Mubu why he had not turned in his rifle at Fort Portal. He explained that he had been offered an immediate ride from town and had not had time to turn it in. Instead, he had turned it in to the police station at Kasese, a town along the main road south. But a check there revealed that they had no record of such a transaction.

An informant at Kanyawara told Tom that Mubu had boasted that he would bribe the police at Kasese to vouch for him having turned in his rifle. Another game guard revealed that not only had Mubu taken his rifle home on leave, but Makuru had slipped away to join him for a few days. Allegedly they shot three hippos illegally and sold the meat. They had even contemplated inviting Mukasa into their business venture but had decided against it because Mukasa was a bad shot. Back in Kibale Forest, Mubu and Makuru were said to have bungled an attempt at poaching a buffalo by allowing the wounded animal to escape.

The Game Department investigated Mubu's case and found no one to vouch for him having turned in his rifle at any police station during his leave. Mubu was dismissed from the department and Makuru was transferred away from Kanyawara to another reserve (a transfer was a common way of "solving" problems). Nyakana found himself without a job. Seven months later wily and wiry Mukasa resigned from his four-year position as game guard. In his opinion it was not a good job. Thus ended the American *Chui* Battalion.

— 12 —

Single Parents

R. P. had drawn my attention to this *natalensis* during a census walk. It was now early February and still the driest period I would witness at Ngogo. The chimpanzees had been hard to find. I was sure this *natalensis* would offset that problem, but my past several nights of sweating fevers, shaking chills, and feeling at the mercy of scurrying trap-wise rats had left me so drained for the past two days that I had not even entered the forest to check the tree. But I had not come to Uganda just to sit in camp and fantasize about committing a mass murder of rats with a shotgun. So this morning I had staked out the *natalensis*.

Something was obviously wrong, but I had taken my daily dose of paludrin, an antimalarial prophylactic with few side effects, every day without fail since I had switched from taking chloroquin six months earlier. So it could not be malaria. . . .

I felt cold. I looked away from the empty fig tree toward the gray sky. The forest was never cold. I pulled my thermometer from my pack. It read 78° F.

I was freezing. I stood and flexed my knees several times to get my blood circulating to my numb fingers and toes. My teeth chattered and my knees knocked. During ten winters in the high Sierra Nevada I had never felt this cold. Suddenly I realized the paludrin must have failed. I had malaria. And camp was two miles away. My head pounded as I started walking.

The trail had never seemed so long before. Each fallen tree across it was a challenge to surmount. I knew the locations of all of them and I began to hate each one in advance. Soon my chills passed and I felt warm. Then hot. Sweat broke from my pores, then stopped abruptly. And I grew even hotter. As my temperature soared I started to worry.

After a half hour I had to stop and rest. I felt a vague disappointment with myself because I was so exhausted. I had not even covered half the distance to camp and I was feeling distinctly worse. I resumed walking but had to stop frequently to lean against a tree. Another half hour exhausted me utterly. I dropped onto the trail to lie on my back. I felt a perverse pleasure in just giving up, in simply lying there too weak to twitch and staring up past the foliage into the little islands of sky. I also felt hotter than I had ever felt before.

Thoughts of a hyena nibbling my face during the night, and of a forest cobra curling up on my warm chest, and of army ants using me for a bivouac made the mud hut up on the hill seem important again. But it was so much easier to lie here than to walk on.

Some time later I did try, and soon the big hill of camp loomed before me. I looked at it hatefully. If only I could sweat. One foot in front of the other. One foot in front. . . . The hill seemed unbelievably steep. Part way up I lost track of what I was doing, but I caught myself, however, still plodding at a snail's pace on the correct game trail. Finally I made it to the door. I felt as hot as a rifle barrel after emptying a quick magazine.

As if in a dream, I mechanically took off my clothes, laid back on my bed, opened the shutters next to me, wetted my bandana with water from my canteen, then placed it on my head. I imagined that steam hissed off it.

This was an acute attack of malaria, the worst among all those I had later, and I had heated myself even more by climbing out of the forest. Because my sweating mechanism had failed, nothing in my physiology was working to curb the high temperatures that cause brain damage, even death. Already I found it hard to think at all; my thoughts were a nebulous spiral of half-formed visions that vanished before coming into focus.

Chloroquin was essential now to kill the plasmodia exploding my red blood cells, but I had given all my chloroquin to Yongili months ago for the Ngogo crew in case one of them had an attack. A month ago Yongili had gone on an eleven-day leave, but I was sure he had returned yesterday. Eventually he would come by here. Meanwhile, to reassure myself that I had not already lost too many brain cells, I tried mentally to multiply 120 by 1,100, but I drifted into delirium and forgot everything.

Hours passed. Yongili did come by, then he returned with chloroquin. The directions were difficult to read. I took four tablets and some aspirin. My head felt as though someone had sheathed a stone axe in it. For several hours I drifted in an aching nightmare of heat. Periodically I emerged to rewet my bandana, replace it on my head, and try to multiply 120 by 1,100. But each time I drifted away before tackling the task that would prove my brain was not fried.

Darkness fell. Chills followed. I thought I would freeze. Maybe this meant the cycle was over, the chloroquin was working. But slowly I heated to fever stage, became delirious, and failed at my multiplication problem again. This was like being possessed. Possessed by protozoa. Around midnight, I guess, the fever broke again. Was the chloroquin killing the merozoites? I took two more tablets. Rats scurried through the house, then through my delirium. But by dawn the fever was gone for good. I owed a large debt to the unknown Amazonian shaman who discovered that chinchona bark would combat malaria.

For a week I felt utterly drained. Chilblains sent irritating pinpricks through my hands, feet, and testicles. It galled me not to be able to search for the apes, even though the dry season was making my searches yield so little. A week of convalescence gave me too much time to think. And, as usual, I thought about my son, Conan, to whom I had been a weekend father for most of his life, except for the past seventy weekends. I had written to him at least as often as I got to Fort Portal and the post office, but he had never written back. I told myself eight years old was too young to be writing letters—I never wrote a letter at that age. But I knew he was mad at me for going to Africa and staying so long. I could hardly blame him. This separation was the toughest part of my research. As long as I was on the job, though, and making progress, it was easier to justify. Now I had too much time to wonder how much he was changing, to wonder how many times he needed me and I was not there. To keep busy, to combat the guilt, I worked on the final chapters of a novel I was writing, *Expedition to Eden,* speculating on what it would be like to explore the origin of man via a time machine. I needed a time machine.

Meanwhile, back in the United States, President Jimmy Carter had introduced three bills to Congress in protest at Uganda's internal slaughter: to prohibit the training of Ugandans in the U.S.A. in any war-related field, to conduct a national boycott of Ugandan coffee, and to expel all the Ugandan Embassies from the United States (the U.S.A. had no embassy in Uganda). Amin took these as a personal, as well as national, affront and tension toward Americans grew. The American State Department was now advising Americans to leave Uganda before it was too late. Most of the eighty-five or so Americans here were missionaries or pilots; none of the former deserted their posts. And, although Uganda had given me some bad moments, I was not ready to leave either.

A State Department advisory done on a local typewriter, which Otim had brought me from my friends the Holy Cross Sisters in Fort Portal (who had been contacted by the Embassy of the United Kingdom acting for the nonexistent American Consulate), directed every American to: 1. maintain close communication with his contact (I had no contact), 2. listen to domestic and international broadcasts (I had no radio), 3. maintain a two-week supply of basic foodstuffs, water, medicines, and gasoline (gasoline was unavailable), and 4. be prepared for travel by having all necessary documents (which I *always* carried in my pocket), valuables, clothing, blankets, food, radio, et cetera immediately ready. As during the previous crisis of Idi Amin Dada versus the Americans of exactly one year before, Tom and Lysa were outside the country.

So, not only was I stuck in camp, I was stuck in a hostile country again. Yongili and the rest of the crew were fatalistic as usual: either all this would pass and be behind us or it would not. The only intelligent course was to lie low and wait for the Man (Amin) to go (be killed or usurped). One day during my convalescence, however, Yongili's equanimity was shaken by a different low source, and he brought it back to camp.

"Where did you get *that?*"

"Benedicto and I were going up the hill to that other forest to cut poles to fix the roof. I stepped on it." Yongili held up a stick with a snake dangling from it.

"Barefooted?"

"Yes, but I jumped away before it could kill me. Then I found this stick to beat it."

"It's called a puff adder," I told him, remembering the Mursi boy at the Omo-Mui confluence whose arm had been such a mess (Chapter 8). Was he still alive? "You know, because these snakes are so hard to see, sometimes almost invisible, they never try to escape or threaten, they just lie there. They get stepped on so much that they kill more people than any other snake in Africa." I bent down to inspect the twitching serpent.

"Watch out for the tail!" Yongili warned me.

"Why?"

"That is how they kill you," he explained straight-faced. "I was lucky it did not hit me."

"Do you know about the fangs?"

"What are fangs?" Yongili asked, pronouncing "fangs" obviously for the first time.

"Teeth, sharp ones."

"What about them?" Knowing Yongili this long, I saw in him a slight resistance building in anticipation of me telling him some new but unbelievable piece of information.

I grabbed the reptile behind the head and it gaped as if determined to expire with an even score card. I gripped it at the corners of the jaw to hold them open, then used a pen to unsheathe the curved fangs.

"See this? It is hollow like a needle. And see this bulge here? This is the poison sack. When a snake bites an animal . . . or a person, little muscles squeeze this sack and force the poison through the fang and into the wound. It is exactly the same as when they give you an injection of chloroquin or penicillin at the dispensary."

"But what about the tail?" Yongili asked, still sure that I was missing the point.

I picked up the tail. "Nothing. You can look. There is nothing on the tail."

"Our elders in Bamba have told us that the tail is the part that kills. Are you telling me that our elders are wrong?"

"Well . . . yes. The fangs do the killing. The tail is harmless—unless you step on it and make the snake angry."

"I cannot believe that our elders are wrong." Yongili concluded. I think he was impressed by the fangs but felt my attitude was too iconoclastic.

I pounded its head with a rock to make sure that it was completely dead, then started to skin it. I hated to see anything die and be wasted, so I was going to convert this fatal misunderstanding to a couple of ounces of needed protein. Yongili watched for a moment, became bored, and observed, "You know, Amin cannot be killed." (He was picking up an unfinished conversation of ours from yesterday.)

"Why not? I asked. "He is only a man."

"But every man who tried to kill him has failed," Yongili pointed out. "Amin has magic. In fact he got it from Mobutu Sese Seko [President of Zaire]. That man is very rich. He can get anything."

"But can't someone else get stronger magic and be able to kill him?"

"I don't think so," Yongili said earnestly. "It would be very expensive. And no *mafuta mingi* [literally "a lot of oil," but slang for a rich man] in Uganda has that much money."

"Do you really think it is only magic that protects Amin?" I asked.

"Of course. African medicine is very strong." He looked reflective. "You know Benedicto's sister?"

"Aboki? The one who always is wiggling?"

"Yes," Yongili said. "She was a prostitute in Fort Portal. And one man wanted her to stay at his home. But she refused. He became angry and paid an African doctor to give her medicine in her tea without her seeing it. The medicine of African doctors, these Bakonjo, is very strong."

"What happened to her?" I asked.

"She became insane."

"How do you know she was really insane?" Actually she did seem crazy to me, but I was playing devil's advocate.

"Man," he answered, quite animated, "she was not reasoning. One time I was there [in Bigodi] and she was going around with no clothes on!"

"And now?"

"Benedicto's family finally saved fifteen hundred shillings from selling *waragi* [bootleg rotgut banana gin]. They had to pay the same African doctor for a cure."

"But have they given it to her?" I asked.

"Of course," Yongili looked at me strangely. "She is sane now."

"How do you know for sure?"

"Because she is reasoning now," he explained as if I were backward. Yongili looked reflective for a moment, then asked. "And you know Nyabihanga?

"That porter who worked for us here recently?"

"Yes. He borrowed many shillings from a *mufuta mingi* in Bigodi but then refused to pay him back."

"Why did he refuse?" I asked.

"He could not pay. He had no shillings."

"So, what happened?"

"The *mafuta mingi* paid an African doctor to send some demons," Yongili explained slowly. "One night they came to Nyabihanga's home."

"What did they do?"

"They killed him."

I could not argue about him dying but neither could I help but wonder what the demons looked like, and whether *they* had shillings. It took me a while, but eventually I realized that nearly everyone here did not simply believe in magic, they lived it. But magic or not, the precarious status of Americans in Uganda worried me again about being able to continue my research.

Unlike in February of 1977, by now I had a lot of data on several facets of chimpanzee life. In fact, many of the patterns I would eventually verify statistically were apparent already. But at this point I was unsure that I had enough data for valid tests. Two months ago, in December, when I had done my one-year progress report, it had been plain that I needed more data on nearly all behaviors, especially ones relating to ecology. But beyond routine behaviors, I was constantly hoping to witness rare ones that would provide more insight into the lives of the apes, especially on the question of their territoriality at Ngogo. Leaving Uganda now would be premature.

Finally, after being restricted to the mud hut for a week that seemed thirty days long, I entered the forest to spend a day at a

Pseudospondias. Blondie, a light-haired, middle-aged, handsome matron with two immature offspring, arrived to forage on its fruit. Blondie was probably the worst mother at Ngogo. She also avoided me as if I were the devil. I had first identified her a year earlier when her tiny son looked like he was all head with a token reminder of a body. Even then Blondie often had let him dangle unattended high above the forest floor while she foraged or groomed with her juvenile daughter, Bess. Blondie strayed from her tiny son farther than I ever saw another chimpanzee mother separate herself from her infant. I named him Butch, because I figured he would have to be tough to survive.

One day while doing a census I discovered just how tough. I found Blondie, Butch, and Bess in a *Ficus natalensis*, but Blondie spotted me only ten minutes after my arrival. Despite my being more than fifty yards away and acting interested in other things, she bolted to exit a tall, branchless, cylindrical bole extending twenty yards from its lowest limb to the ground. Blondie and then Bess handled this easily, but Blondie had abandoned Butch. To catch up with his mother, he descended to a point about thirty feet up, where the bole became too wide for him to reach around and grip.

He seemed stuck. It looked as though he needed longer arms to descend farther. Butch whimpered in rapid, ululating notes I had never heard before. Then he inched down farther holding on by the pressure of his fingers against the rough bark. As it became even more difficult for him to grip, he cried out in raspy, single notes, about thirty of them in the next twenty seconds. Neither his mother nor sister returned to assist him. After a full minute of crying and descending, he had inched into the understory about fifteen to twenty feet above the ground. Then he lost his grip and fell. I heard him thud against the earth. Although he was hidden, I could hear him, and my heart ached for him.

I was disgusted with Blondie for being such a selfish mother, and I was worried that Butch had been injured in the fall. But I did not dare to approach closer to investigate because I was worried about panicking Blondie into abandoning her son completely. But two days later I found them again at the same *natalensis*. From concealment I studied Butch for three hours as he played, suckled, and simply dangled from limbs. He seemed uninjured. Butch *was* tough.

As I watched them now, Butch excavated a deep knot hole in a crotch of the *Pseudospondias* bole. He industriously dragged out a pile of soggy leaves and strew them about, some into his complacent mother's lap. Once the hole was cleaned out, Butch dipped his hand into it, then pulled his hand out, held it aloft, and watched the droplets dribble down his arm. Before they reached his elbow he caught them on his tongue. Then he jammed his hand into the hole again to repeat the whole process.

Too soon, though, Blondie spotted me. She stared at me for a couple of seconds as if unsure of what I was, then abruptly climbed down, again leaving Butch behind. As if an afterthought, she paused and turned to face him. He clambered down to her and climbed on her back, then they vanished.

Several other mothers at Ngogo—Ardith, Clovis, Farkle, and Mom, for instance—seemed exemplary in their constant awareness of and concern for their infants. Chimpanzees face few enemies other than man and parasites (and other chimpanzees), but the few predators they do face threaten young chimpanzees most. Ursula Rahm reported a young infant killed by a leopard that first attacked its mother, despite the apparent counterattack by a male chimpanzee. Adriaan Kortlandt (1962) conducted experiments presenting a stuffed leopard both to free-living and captive, wild-caught chimpanzees in Guinea, several of whom charged it with sticks, while a few even clubbed it. Wild chimpanzees in Mahale actually attacked leopard cubs in a cave. Geza Teleki (1973b) reported a Gombe male who stamped and slapped a serval cat that had been wounded first by baboons. I found no conclusive evidence for leopards around Ngogo, but hyenas did hunt in the forest. But without doubt African crowned eagles pose real danger to small infants separated from their mothers in open sections of tree crowns. These raptors not only specialize in primates, they are capable of killing *large* primates—they have been observed killing adult colobus, baboons, and even human children. Only the mother of an infant chimpanzee could prevent it from being taken. All in all, however, the frequency of reports of chimpanzees being preyed upon by carnivores or raptors is extremely rare. More significant may be the danger posed by other chimpanzees.

As mentioned briefly in Chapter 10, several independent researchers in three separate study areas in Tanzania and Uganda have witnessed nearly ten episodes in which adult male chimpanzees have brutally murdered infants—primarily of females apparently unfamiliar to them. Males of Mahale's M-Group also have killed infants of females who were not strangers but merely resided mainly in peripheral areas of the community's range. Human sensibilities assume these killings must constitute the darkest side of chimpanzee social life, but they do not. Sociobiologically, these males had good genetic reasons to kill them: the infants (with possibly a few exceptions) were offspring of alien males and, by killing them, the males both eliminated future competitors and prevented them from remaining in the males' territory and thus intensifying competition for limited resources (it is revealing that, in the 1985 summary by Toshisada Nishida and Kenji Kawanaka of all infanticides, adult males have murdered male infants eight to one over female infants). A second reproductive advantage from killing remains potential. Once a mother loses her infant, her estrous cycle resumes and she may mate and conceive a new infant within a month or two—instead of after several years of nursing. So, from the perspective of Darwinian selection, the killings make sense (see also Chapters 1, 10, and 14).

Similar reports of within-species infanticide by males and/or females, as part of an apparent strategy to increase the killer's reproductive success at the expense of that of the parents of the murdered offspring, have been reported for many other vertebrate species, including many fish, amphibians, birds, rodents, carnivores (lions, tigers, pumas, cheetahs, brown bears, coyotes, wolves, wild dogs, brown hyenas, and others), plus several primates (red howler monkeys, Hanuman langurs, savanna and chacma baboons, mountain gorillas, and *three* other species in Kibale Forest: redtails, red colobus, and blue monkeys), and, of course, among humans. Distressing and distasteful as it is, infanticide is a widespread phenomenon (and it is analyzed particularly well in the book *Infanticide*, edited by Glenn Hausfater and Sarah Blaffer-Hrdy). These killings by chimpanzees *were* ghastly, but they did make sense, and they were not the worst of what chimpanzees are capable of doing.

Again the example comes from Gombe. Jane Goodall (1977, 1979, 1986) reported an adult female, Passion, who attacked another mother, seized her infant, then ate it, sharing the kill

209

with her own two offspring. One of these, an adolescent female named Pom, grabbed the next infant of the same victimized mother a year later and again Passion's family ate it. A month later Pom attacked another new mother and murdered and ate her infant. Goodall reported that during the three years between 1974 and 1976, only *one* infant in the entire Kasakela Community survived beyond the age of one month. Six others vanished and are suspected victims of cannibalism by Passion and Pom.

During ensuing years new mothers steered clear of Passion and Pom, and often were accompanied by an adult male who defended their infants. Sadly, as bizarre as they seem, these murders also made some sociobiological sense for Passion and Pom, who gained mammalian prey normally very difficult for females to capture, and who lowered competition their family would face for other resources. But unlike the killing of alien infants by males, these killings by Passion and Pom are obvious exceptions to the general social mores of chimpanzee society, and, if a code of law existed, they would be considered crimes. Hopefully, these murders by Passion and Pom *do* constitute the worst of what chimpanzees are capable of doing. They also reveal the challenges a chimpanzee mother faces from unexpected quarters.

Ardith was an Ngogo mother who impressed me as exemplary. She normally traveled with her two youngest sons: Ashly and Anson. Socially Ardith seemed on equal terms with all the other adults with whom I saw her. She and her sons appeared healthy. What else was there?

Why Ardith appeared so exemplary was not always clear to me, but, during my first year at Ngogo, I began to take it for granted. So I was shocked when I started seeing her now, in February, in estrus and traveling alone. I held onto a glimmer of hope that Anson was traveling with Ashly, which for an infant consistently to do while his mother was alive, however, seemed very unlikely. But my hope was proved false again and again as I saw Ashly traveling only with R.P. Finally, after months of seeing Ardith with Ashly but not her infant Anson, I gave him up for dead. But every time I saw Ardith traveling alone I could not help but wonder what had happened to little Anson, and wonder whether his demise was due to a lapse on her part or to circumstances

210

beyond her control. How hard is it to be a good chimpanzee mother?

It is difficult to think of another word more subjectively loaded than "motherhood." Everyone has had a mother, good, bad, poor, or indifferent. And everyone has some notion about exactly what it is that makes a mother good or bad. But sociobiologically, a mother must pass only one test—a genetic one—and her score on it determines her final grade in the natural field course of evolution. That score is Darwinian and consists of the number of offspring she produces who survive to adulthood and themselves become healthy parents. In a sense, her final score is measured in grandchildren. But chimpanzee mothers face real difficulty in achieving a high score because they are invariably single parents who shoulder the entire job of raising infants themselves— without childcare centers, welfare, food stamps, ERA, PTA, or grandma's help—and with no help from a mate.

A successful chimpanzee mother must be a good mother, or a lucky one, because none of her shortcomings can be repaired at a clinic. Unlike most species of animals, however, among chimpanzees good mothers are not simply born that way. In fact, unless they are trained through example by their own or other good mothers, even potentially excellent chimpanzee mothers will grow up to be totally inept. Nancy Nicholson found that infants captured in the wild at age two did reasonably well upon becoming mothers a decade later in captivity, which suggests that a female's first few years of life are critical in learning maternal skills and perhaps the maternal attitude.

Being a single parent myself probably influenced my subjective judgments about how good each Ngogo mother was. I found myself feeling inordinately proud of the good ones. Clovis and Mom were two of my instructors. I found Clovis one day carrying Chita, her two-year-old male son, through the foliage of a large *mucuso* as she foraged. Her older son, Clark, was with her as usual but foraged on the opposite side of the crown.

Chita sat in her lap as she routinely munched figs. After studying her a while, he waited until his mother had picked a fig and was holding it, then he pulled her right arm toward him and ate part of the fig from her hand. Then he released her arm and dropped fragments of the fig. His mother held it in the same position for another ten seconds or so, but Chita squirmed and showed no further interest in it. Eventually Clovis ate it herself.

Sharing food was rare among chimpanzees in Kibale; it was rare among chimps observed everywhere. Joan Silk investigated food sharing among Gombe chimpanzees and among captive chimps at Stanford University. While it was common for a mother to share certain foods with her infant when the infant solicited it, *un*solicited food sharing was so rare as to be considered nonexistent. But after I observed Clovis sharing with Chita, I witnessed apparently unsolicited sharing of food by another chimpanzee mother at Ngogo.

She was a "new" chimpanzee whom I had named Mom because she was so good with her infant, Munch, who looked less than a year old and clung to Mom tenaciously. In the midst of climbing, scanning, and picking fruit, Mom paused, picked a fig, and placed it in her infant's mouth. Then she picked another and ate it herself as she climbed. I watched fragments of fig drop from Munch's mouth as she chewed. She clung to her mother ventrally during all my observations and never picked a fig on her own, but neither did she beg by gesture or vocally as Mom repeatedly supplied her with fruit.

A mother's sharing food with her unweaned infant is sometimes sensible because she is providing food to her infant in the form of fruit that otherwise she would have to provide as milk during suckling. Because the metabolic conversion of fruit to milk in her body is less than 100 percent efficient, such food sharing seems an economical shortcut. Possibly, though, this shortcut would harm an infant not yet ready to process fruit on its own. Specifically, any reduction of milk intake, a more perfect food with built-in immune agents, might delay the infant's ability to become self-sufficient in its foraging. Somehow a good mother must achieve this balance between providing milk and providing an education.

Nutritional questions on the diets of chimpanzees are a long way from being answered (Chapter 4), but we do know that captive chimps reared on diets in primate research centers and zoos attain adult weights 50 percent larger than those reported by Jane Goodall (1983) for Gombe chimps. A substantial part of this extra weight is structural, not fat, and provides strong evidence that nutrition and the availability of food is a *valid* preoccupation among wild chimpanzees. Because of seasonal vagaries in the availability of food, optimum nutrition is never maintained for long. But because good nutrition plays such an important role in each ape's development, adult size (and hence status), and reproductive physiology, it is easy to understand the intense focus chimpanzees display on food.

A mother's sharing of food with her infant probably is of minor nutritional advantage to it until it is older, but often when it is old enough to be satisfied by such foods, it has the ability to harvest them on its own. So why do some mothers share? Sharing foods helps communicate to the infant which parts are edible. Joan Silk observed Gombe mothers responding differentially to their infants' solicitations for food: they most often shared with them foods difficult, rather than easy, to process. Silk concluded that sharing probably promotes autonomy in the infant that will reduce the trauma that comes with its eventual weaning.

Weaning is a rough hurdle for infants for more reasons than one. Cathleen Clark reported that Gombe mothers sometimes begin weaning their infants during their second year but normally do not complete the process until age four. The end of weaning (which usually accompanies imminent pregnancy in the mother) induces a classic depressive syndrome in the infant: loss of appetite, a lessening of play with other infants, frequent huddling for long periods, and increased insistence on physical contact with its mother. This depression may last a few months or up to a year and may include increasing demands on the mother, tantrums, and a regression to a more infantile behavior. As do the attempts by infants to prevent mothers from mating (Chapter 10), these profound changes prompt an important question: what do infants hope to gain by acting this way? *Is* there something to be gained?

Parental investment. This key concept was defined by Robert Trivers in 1972 as "any investment by the parent in an individual offspring that increases the offspring's chance of surviving (and hence reproductive success) at the cost of the parent's ability to invest in other offspring." When viewed in evolutionary terms, conflict between a mother and her infant over the amount of parental investment she provides is automatic. The explanation is genetic, or, in other words, sociobiological. An infant is genetically related to itself by 100 percent, to its mother by 50 percent, and to its next sibling (who likely will be sired by a different father) by only 25 percent. When an infant reaches the age when its mother decides, based on her *equal* relationship (50 percent) to each of her offspring, that it is time to invest in a new infant, the genetic gain to the *old* infant is much lower than it is to the mother, whereas any *continued* investment in itself the old infant can coerce from its mother that will better its chances of survival will be of higher genetic value to it than starting that new sibling. Of course, the old infant eventually will reach an age when its survival is more assured and continued parental investment is more negligible, at which time its genetic interests will coincide through inclusive fitness with its mother's decision to have another infant. From a selfish perspective the infant *should* coerce its mother with tantrums and demands to keep nursing it, instead of getting pregnant, until it reaches this older, safer age. But the mother's reproductive lifespan is limited to only four or five offspring who survive to at least five years old, and *she* is programmed by her own genes to produce the maximum number of healthy ones she can, tantrums or no tantrums.

A female's choice of apportionment of parental investment in each of her children must be balanced carefully to maximize her reproductive success. Because in the long run natural selection favors *only* those mothers who do so, maximizing reproductive success is *the* critical part of being a good chimpanzee mother. Based on this Darwinian standard, as well as the survival and health of her own offspring, old Farkle, with at least three daughters (Fanny, Fern, and Felony) and a probable son verging on adulthood (Fearless), was the best mother at Ngogo. And, apparently, she accomplished her success without help.

Being a single mother in the sexist world of chimpanzees seems unfair. What about parental investment by males? Again the weak link is ecology, though the whole story is more complicated. During lean seasons when food is limited to small, hard-to-find patches, a hungry male is the last thing a chimpanzee mother would find beneficial in raising her offspring. She would be forced to compete with her more dominant partner for any food they found. Because predators of chimpanzees are rare in most habitats, it probably is a good gamble for females to assume the entire burden of parenting and thereby avoid that competition for food. And, despite what appears to be a feckless existence by adult males, they do provide some parental investment. In fact, the concept of parental investment itself is indispensable in understanding chimpanzees as naturally occurring social killer apes.

In social systems where parental investment is markedly un-equal between the sexes, the sex making the far greater invest-ment, in this case females, becomes a limited resource for the other sex. This is why: female chimps invest the entirety of their adult lives raising a few offspring. No strategy exists by which they could innovate and do better, and thus outcompete conven-tional females. A male, however, could inseminate a new female every day if he could find the opportunities, or, more to the point, the estrous females. And, according to the Darwinian logic of sociobiology, males should try to find those opportunities. But because approximately only one female exists for each male, females are the most limiting resource to each male's reproduc-tive success. In most species in which females contribute far more parental investment, males are free to compete fiercely with one another to gain additional opportunities to mate by excluding male adversaries. This competition leads to natural selection among males (Darwin termed pressures caused by the reproductive contest between the members of the same sex of a species "sexual selection") for improved hereditary weaponry: greater size, speed, strength, intelligence, aggressiveness, the pro-pensity to form advantageous alliances, and a demeanor that attracts females.

Most species of mammals fit into the category of unequal parental investment, and aggressive male primates of many spe-cies use the common strategy of mastering a harem and zealously defending its territory against male intruders. In such social sys-tems one male is pitted against another and individual prowess decides the issue. If the data from Gombe and Mahale are cor-rect, plus everything so far from Kibale and other studies, male chimpanzees, as absentee fathers with plenty of time on their hands to seek out additional opportunities to mate, have taken their competition for mates a quantum evolutionary leap beyond this common primate system. In their contest against other males to monopolize females, male chimps somehow have revolu-tionized normal primate sociality by retaining their sons and grandsons as allies. Instead of individual males competing, sexual selection appears to have coalesced chimpanzee males into male *kin groups*. Against such kin groups, who *share* reproductive success through inclusive fitness, solitary males would have no chance in the reproductive contest.

This explanation, or sexual selection, model was in harmony with all I had seen in Kibale Forest. Tight solidarity between males at Ngogo and at Kanyawara was blatantly obvious. But I had still seen nothing so far to indicate either of the two major consequences of a chimpanzee social structure based on male kin groups: strict territoriality and combat between males of different kin groups. The following two chapters explore and discuss sexual selection, the key concept (along with inclusive fitness) to understanding the reproductive logic of male cooperation. Chapter 13 explores the ecology of lions living south of Kibale and the reproductive strategy of males, which has an unusual and revealing resemblance to that of chimpanzees. Chapter 14 returns to sexual selection and inclusive fitness among male chimps, discusses their reproductive strategy and parental investment and, using my final observations at Ngogo, sums up the roles of both sexes in the chimpanzee community.

—— 13 ——

Ape Man Meets Lion Man

"**L**ions have killed a warthog just down the hill!" a Ugandan announced to Karl as we prepared for a sojourn in Ishasha. "Where?" Karl asked, not quite believing. Karl Van Orsdol, a Californian of German descent attending a British university (Cambridge) and conducting his Ph.D. research on the predatory patterns of lions in Uganda, was Struhsaker's other student advisee. During his first month of research (March 1977) Karl and I had decided on an exchange program wherein each of us would spend several days on the other's project to broaden our horizons. Despite only one hundred miles separating us, a full year had passed between the beginning of his work and our managing phase one: my visit to his project. Though I was late in finding the time to join him (Amin had finally relaxed his anti-American stance again), diverting a week from Ngogo proved definitely worthwhile; I was about to discover some unique parallels between lion and chimpanzee social structure that would make this visit far more valuable than I had anticipated.

"Just here, down the hill," he said, pointing vaguely along the Mweya Peninsula, an islandlike landmass jutting into Lake Edward like a swollen thumb.

217

We ran to the edge of the plateau and scanned along the eroded, brush-dotted slope leading to the flat terrain below. No lions were visible.

"Where?" Karl asked again.

"Just down there; you can't see them now," a second man volunteered. He was older and had a paunch.

No possibility existed that Karl would leave this report uninvestigated. He gathered his notes and identification file and got into my Land Rover. A third neighbor, a young woman named Peace, asked me if she could come along. Both men waited expectantly. I said she would have to ask Karl.

I could see Karl imagining what would happen if they came. If we found lions, we would stay to observe them, possibly for twenty-four hours. Soon the Ugandans would become bored and want to return to headquarters, while Karl and I would not want to leave and risk losing the lions. Interrupting observations would be a certainty. This conflict of interests would lead to an unpleasant stay or interrupted observations.

Before Karl could explain diplomatically why we could not take all of them along, Peace wagged her finger at him, "Here we have come to you as good neighbors to help you with your work, and you refuse to give us a ride, when, as anyone can see, there is plenty of room in the Land Rover. And, besides, we will be quiet."

Soon the five of us were cruising across the gullied slope, pausing to peer into each clump of brush, and constantly doubling back to circle around gullies too deep to cross. Peace sat between Karl and me, and the two men sat in the heat of the covered bed and ate our bananas. The paunchy one offered an uninterrupted stream of lion lore and pointed out bushes where they might be.

"I have eight children and all of them are sons," Peace announced to me after fifteen minutes of unsuccessful searching.

Because she was only nineteen years old, I secretly doubted her veracity, but I answered, "That's a lot of children." (Later she admitted that she had only three.)

"Do you know what tribe I am from?"

Peace was healthy and attractive despite her alleged brood of eight, but she had no facial scars, and the tight braids along her crown were an insufficient clue to her tribe. Ankole cattle herders dominated this part of Uganda.

"Ankole," I said.

"How did you know?"

"I'm not as dumb as I look."

She pondered this for a moment and then, almost accusingly, asked, "Where is your wife?"

"I have no wife."

Her brow furrowed and she was about to reprove me for leading her on, when Karl came to my rescue. "He is not Tom Struhsaker; he is another American."

"Oh," she said. Then I could almost hear her mental gears turning. "You have no woman?"

"No."

"Are you normal?" she asked.

"I used to be. I guess I still am. . . ."

She seemed relieved. "Good. I will send you my sister tonight."

This was one more item for contemplation as I wrestled with the steering wheel to avoid gullies, boulders, and thickets. The noon sun was oppressive and the vehicle poorly ventilated. Karl asked the paunchy one again where he had last seen the lions.

"I'm not sure exactly where, but I think I can find them." Abruptly he climbed out through the tailgate to peer into the thickets at close range. He was defenseless against any lion who took offense at his overfamiliarity. Only three months earlier, a lioness had attacked three people beyond the northern boundary of the Queen Elizabeth Park. His danger was real. I voiced my misgivings and Karl agreed, but the second man in the bed explained, "He does not fear lions. He practices witchcraft and they will not hurt him."

The paunchy shaman wandered from one thicket to another but found nothing. Finally he hiked back up the slope to where he had originally seen the lions, relocated the correct thicket, and directed us to it by signs. In it were two lionesses, one of whom was young (Karl's code name for her was no. 069) and the other an adult (no. 066) who had cubs hidden somewhere nearby. Both were gaunt and did not appear to have made a kill.

Soon the pair abandoned the thicket, then padded a detour around Hugo, a one-tusked bull elephant who routinely made forays to park headquarters to sift through the garbage. Hugo lost his other tusk after goring a man who threw rocks at him to shoo him away. Months later a second man tried the same thing. Hugo picked him up in his trunk and smacked him on the ground. "He looked like he had been run over by a truck," Karl explained. (Hugo continued to prowl headquarters without retribution until January of 1979; during Tanzania's liberation of Uganda from Amin, Tanzanian troops would machine-gun him for his remaining tusk.)

The lions and elephant otherwise ignored one another, but as soon as they passed him they froze to study an approaching warthog. Both great cats took full advantage of natural cover as they moved closer to ambush the warthog from a path intercepting his. Most lion hunts are failures, especially during daylight. This warthog trotted jauntily toward the lions, seemingly oblivious of their ambush, but then veered from his original direction to climb the slope. Both lionesses relaxed visibly, then slowly walked a few hundred yards to lie up in another thicket.

Karl decided our best course was for him to return to the house to finish packing for our trip to his second main study area in Ishasha, while I monitored the predators in the bush. After packing, he would join me to follow them all night, if possible. I dropped Karl, Peace, and her companions off at the houses, then hurried back to the plain. The lionesses did not move until a half hour after I returned, and then they slept until Karl joined me. From fifty to a hundred yards behind, we followed their leisurely pace to the denser thickets to the west. They paid us no attention. At about 5:00 P.M. we followed them into a broad thicket that looked impenetrable. There the older female retrieved her two, six-week-old cubs. They were bundles of playful energy: they pounced on one another, batted their mother's tail, and stalked anything that moved, including the Land Rover.

After about forty minutes, no. 066 stood and walked south for about ten minutes with the cubs; no. 069 followed closely. The family paused on a ridge where both hunters gazed downslope, then padded east. For a few seconds the tiny cubs sat on their haunches to watch them go, then they silently melted into a thicket to vanish. I wondered how they knew what was expected of them.

I hurried to avoid losing the lionesses in the thick bush. A few hundred yards from the concealed cubs I rounded a thicket to see both females frozen side by side in a crouch and staring ahead from the scant cover of a small bush. We were so close that I had to stomp on the brake pedal to stop less than twenty yards behind them. But their concentration was so intense they did not even twitch.

From my position I could see fifty yards beyond the hunters to an African buffalo cow browsing at the edge of a thicket. The cow was unaware of the two cats whose eyes monitored her every move. Both remained frozen, flank against flank, for another full minute. They were statues about to explode.

Number 066 suddenly and rapidly slinked into the open, then froze again halfway to the cow as the latter turned toward her while browsing. Simultaneously no. 069 moved with the older cat but continued past her on a more direct line to the cow and stopped only fifteen yards from her and six to eight yards from no. 066. Both predators crouched flat against the overgrazed plain without even a tussock of grass to hide them from the buffalo. They remained perfectly immobile for the next seven minutes. The bovid turned to face both lionesses, but apparently did not see them; she settled onto her belly to chew her cud. Then she turned away from the predators, but they still made no move. It was as if they had been transformed to stone.

Number 069, then no. 066, abruptly rushed her and she lurched to her feet. A tiny calf appeared behind her and the younger lioness leapt at it and bowled it over. Number 066 reared up to grip the cow along her side, high and forward, but approached from behind. The buffalo tried to shake her loose, and the cat slipped so that her body was dragged along the ground between the cow's legs, but her claws held their grip. As the cow turned to try to gore her, no. 066 stood on her hind legs and hopped repeatedly to remain directly behind the bovid, while never loosening her claws in the cow's rump.

Meanwhile, the younger lioness stood over the prostrate calf and watched with little apparent interest toward her pride mate who continued to dance behind the increasingly more frantic cow. Number 066 retained her grip high on the buffalo's haunches and repeatedly bit her near the base of the tail. All the cow's efforts to dislodge her and spin to face the cat were negated by the latter's well-timed hops. A long bloody strand of what appeared to be placenta emerged and hanged from the cow's perineum.

After two full minutes no. 066 finally lost her grip. The enraged cow immediately pivoted to chase her attacker around the thicket, which was only about fifty feet across. As they finished the first lap, no. 066 ran past no. 069 still crouching over the prostrate calf. The cow veered toward no. 069, who turned to face her. The buffalo charged both lionesses, who wheeled to flee around the thicket again with the bovid in hot pursuit. Halfway around it the older predator dived into it to shortcut and emerge at the calf ahead of the other two. The cow completed her second lap and chased no. 066 away from her calf. But both lionesses spun to face her and simultaneously feinted a charge. The cow turned away from her newborn calf, who had not moved for the past three minutes, and, mooing deeply, loped to the southwest.

The older lioness quickly dived for the calf and clamped her mouth over its muzzle in a suffocation grip. Two minutes later the calf kicked once. The younger predator sat about twenty feet away as if disinterested. After five more minutes, no. 069 walked to where her partner had first attacked the cow, sniffed the ground, then gazed in the direction of the now vanished bovid. The elder cat still held the calf, which kicked again.

The young lioness returned toward the calf but stopped ten feet short. Then she dashed for its hindquarters. Her pride mate spun and snarled, quickly gripped the calf by the neck, dragged it nearly a yard farther away from no. 069, and shoved her own hindquarters toward her instead. The calf bleated once. Despite this tension, both females began feeding on the calf as if the bleat had been a dinner bell.

After a half hour of feeding, the younger female abandoned the carcass and sat patiently nearby for two hours while no. 066 consumed it. When she finally quit the carcass, no. 069 rushed to the scraps and picked through them for only seven minutes.

The mother lion padded back up the slope in the darkness. The younger female soon followed along a zig-zag path to meet no. 066 and her cubs, then all four crossed the ridge. Number 069 played with the cubs as they trotted after their mother double time, batting at them with her paws, but they were so tiny that they quickly rushed away to avoid being bowled over. Soon they found a small thicket and disappeared in it. The young lioness emerged from the other side and collapsed for a nap. An hour later she looked up as a hippo thudded past, then she nodded off to sleep again. After another hour, three more hippos wandered past, and no. 069 glanced at them too. The pace of the evening was lowering. After midnight, so were our eyelids.

Early April was Karl's thirteenth month in Queen Elizabeth National Park, yet the episode described above was only the fifth successful hunt he had seen during daylight and the *first* in which an adult prey had been attacked. By contrast, he had logged over forty successful hunts after nightfall. Long after midnight the savanna became as quiet as the grave, and Karl concluded that a second hunt—or any new activity—was unlikely. We returned to headquarters. Inadvertently I had stood up my first blind date in Uganda.

The southern salient of the Ruwenzori tapers to crater-pocked foothills straddling the equator, then to rolling savanna surrounding Lakes George and Edward. Zaire's large Virunga National Park hugs the west side of Lake Edward, while Queen Elizabeth Park mirrors the east side and extends south along the Ishasha River and northeasterly toward Kibale Forest to surround Lake George. The park is nearly 2,000 feet lower than Kibale Forest and drier; it fits the Westerner's image of African wilderness: predominant savanna dotted with acacias and giant euphorbias and cropped by thousands of buffaloes, Uganda kob, topis, hippos, warthogs, waterbucks, and elephants—who in turn are preyed upon by lions, leopards, and hyenas.

Karl was based at the Uganda Institute of Ecology at park headquarters on the narrowest part of the neck of Mweya Peninsula, and next to a modern but vacant tourist lodge. Because of Amin's eradication of the tourist trade (officially resumed in 1975, but in fact dead), the spacious lodge, with its empty storerooms and nonfunctional staff, seemed surrealistic, a living museum of colonial times.

Late that morning we loaded both Land Rovers and headed east along the roundabout drive that would take us south to Ishasha, the southern end of the park and Karl's second study site. We needed both vehicles because I would return to Ngogo a week before Karl would return to Mweya. As we descended the hill from headquarters we passed two Ankole women trudging upslope, bent under loads of valuable firewood for cooking gathered from the bush and suspended by tumplines across their foreheads. Thousands of voracious hippos had grazed the grass everywhere, even with the stony soil now gullied by erosion. These natural lawnmowers were east along the bank of Kazinga Channel and west along the curving lakeshore, where a fishing canoe hove into view. (When designated a park, Queen Elizabeth included nineteen fishing villages within its boundaries.)

The sternman propelled the two-man craft with a heart-shaped paddle across the turbid water and past a dozen motionless hippos betraying their presence only by nostrils and eyes protruding above the surface. The second man, in the upswept bow, reached into the bilge for a buoy with a leafy limb for a flag. He set it in the water, then fed about fifty yards of net into the lake as his partner guided the canoe in a semicircular arc facing the shore. They continued toward shore, then faced their net, and the bowman lifted a long pole and slapped it resoundingly across the gulf. Hippos increased their distance beyond the net. As the sternman maneuvered them into the arc the bowman slapped his pole against the water to frighten tilapia through the murky depths to be trapped by their gills in the waiting net.

The road south was the main route into Zaire and saw a constant traffic in heavy trucks supplying the eastern fringe of this vast forest nation, whose far-flung provinces were only lightly touched by the authority of Mobutu Sese Seko. But this highway into the region Stanley called "Darkest Africa" was only a narrow dirt road, which during rains became so rutted and mired as to defy description.

This road lived up to my expectations. The southern stretch through Maramagambo Forest consisted of two deep ruts in a ribbon of semiliquid mud many miles long. I marveled that any truck could traverse it. In fact, most truckers had parked their rigs just north of it at a tiny *duka* (store) to wait for a dry spell to make passage possible. Other drivers not as wise were now hopelessly mired along a slope. I looked ahead at the row of semitractors and trailers canted at odd angles, some buried three feet deep into the roadbed, and my stomach sank. A delay looked inevitable.

We stopped and walked alongside the sorry convoy and greeted the truckers, who had faced the inevitable and were cooking dinner over Primus stoves. I found a route. Karl said he would follow if I made it. I did. He did not. So I walked back and put my shoulder to his short Land Rover, then two truckers added theirs, and, amid cheers, we were soon south of the convoy.

After nightfall we arrived at Ishasha Camp and took up residence in one of the huts intended for tourists, a vanishing species in Uganda. After tea, bananas, and groundnuts we drove into the night in Karl's Land Rover, armed with his directional antenna and radio receiver. Karl had radio-collared one female each in the Mweya and Ishasha areas. Originally, before arriving in Uganda, I had considered collaring a couple of chimpanzees in Kibale as an aid to documenting their full range, but lack of funding and my worries about injuring the apes during immobilization squelched that idea. I asked Karl how much help the transmitters were.

"It's not a question of help," he explained. "The transmitter makes it *possible* to find the lions. It's like the difference between trying to read in the dark and trying to read after you have turned on the light."

Now the rolling savanna was shrouded in starless black as dense as tar. Floating red eyes danced before the beam of the headlight: topis, kob, buffaloes, and an occasional hippo. This was the time of their maximum vulnerability; under the cover of darkness hungry predators stalked the unwary.

We stopped on a hilltop in a landscape I had yet to see. Karl set the antenna on the hood and rotated it in search of the beacon from the collared lioness. The receiver said the lioness was either no longer near Ishasha or else was not transmitting. We puzzled over this for a while, then Karl concluded her transmitter was malfunctioning. We were reduced to reading in the dark.

Amazingly, shortly before midnight, Karl found the collared female along with two other lionesses wolfing down a topi calf. They paid us no attention. Between each visual check with the spotlight were dark intervals of fifteen minutes, during which we were silent because our ears were the only links to the predators. Their bellies swelled and our eyes grew heavy. After midnight we found ourselves in the same predicament as the night before: we were too tired to stay awake. The lionesses abandoned the kill and melted into the night. We returned to the hut.

For the next five days I was snared into Karl's Catch-22 schedule. To be sure of observing lions at night when they hunt, it was necessary to find them during the day to learn where they would be in the evening. Too often, though, finding them took most of the day. Ideally, once found, lions should be followed all night until they lie up, after which we should sleep most of the day. But what actually happened was we searched most of the day, observed half the night—until we lost them or were too tired to function—wandered zombie-like back to the hut, then started again too early the next day to last through the night. This did acquire data, but it also drained our stamina. Because Karl had amoebic dysentery (though he did not know it yet), was malnourished, and half-atrophied from spending all his time in a Land Rover, fatigue came quickly. But the fact that we could not stay awake did not stop us.

Finding lions during the day was easy along the Southern Circuit, an itinerary connecting large *Ficus gnaphalocarpa* trees favored by lions as resting places. We merely checked each tree until we hit pay dirt, then, after watching them snooze, we followed them without lights at dusk when they prowled. This sounds simple but it is difficult to stay far enough behind lions so as not to spoil their hunting, yet stay close enough to notice in pitch blackness that they have just disappeared into a thicket. When one considers that a lion's coloration and behavior have been refined by natural selection for the ultimate in ability to crouch undetected in ample light by prey whose survival depends upon avoiding lions, it is easy to imagine how quickly they can be lost in the dark.

During our first day at Ishasha we found a lion and lioness in consort.

"It surprised me at first that males seem to have a sense of propriety over who gets to mate with an estrous lioness," Karl said. "The first lion she lifts her tail for normally becomes her consort—they stay together for a couple of days and mate at least a hundred times."

"Like chimpanzees on 'safari,'" I answered. "The lack of competition between male lions probably results from the same circumstance: cooperation between male kin groups—brothers in arms."

"It makes sense," Karl admitted, "It's definitely not like other mammals."

We lost the consort pair after dark, but soon found a male and two lionesses stalking a topi. They rushed at the antelope. We gave them a few seconds to conclude the hunt, then moved closer. All of them had been swallowed by the night. Additional searches failed.

We got off bleary-eyed the next morning and soon found the consort pair again, now sleeping in a thicket. We moved on and found three Northern Pride lionesses asleep in another thicket. Late that afternoon we spotted two Southern Pride males in a fig tree, but we lost them an hour after nightfall. Despite having done all of that preventative research during the day, further searching revealed scores of red eyes floating in the dark, but none of the green eyes of lions.

Around dawn that morning we found two lions and six lionesses of the Northern Pride, including lioness no. 030 wearing the transmitter Karl decided was faulty. He had a new collar ready, but to replace hers he had to shoot her with a drug dart before the sun became intense or she might overheat and die. And to be sure we were not attacked, she had to be away from other lions. We waited. For an hour a lone female kob stood on the ridge overlooking the lions and whistled in alarm. The pride ignored her. A herd of buffalo moved closer from the west. Karl whispered that it might try to run the pride off, a common maneuver of buffalo here, but, though they stopped and stared intently, they grazed beyond the lions. The sun rose and made shooting no. 030 too risky. We drove up over the ridge and planned to return later. (Eventually Karl discovered that no. 030's collar was okay; it was the receiver that had gone haywire.)

We stopped nearby a kob lek to scan the rolling savanna. Males of this mid-sized antelope develop stout, lyre-shaped horns; females are smaller, hornless, and resemble female deer. In Queen Elizabeth Park and on the Semliki Plains virtually all kob mate on unique display grounds called leks. Helmut Buechner reported that leks averaged about an eighth of a square mile and contained 30 to 40 individual territories up to sixty-five feet in diameter. Adult males repeatedly battled one another for proprietorship of a display territory, which eventual depletion of stamina forced them to lose after about ten days. Across this well-trampled checkerboard of male aggression strolled fickle females in estrus, eliciting elaborate precoital displays from one eager male then another, each of whom ceased displaying as the female crossed into a neighbor's territory.

The clusters of estrous females milling about the periphery here seemed oblivious of the numerous vanquished or immature males who loitered with them. (Buechner found that more than half the adult males belonging to a lek spent all of their time within 550 yards of it, while only 14 percent of the females remained so close.) A female entered the lek. With chin held high and horns all but invisible along his back, the nearest male pranced toward her. After sniffing at the female's perineum, the male curled his upper lip in a grimace called "flemen" to analyze her scent for estrous hormones. Then he circled her closely, as if herding her, and tried to mount. Typically, this female did not allow the male to mount so quickly. Only after several attempts does a female allow mounting, and rarely with the first male who courts her.

While this coy routine occurred on a few scattered territories, nonterritorial males entered the lek to challenge holders of the aphrodisiacal ground. Only full-grown adults had a chance of usurping another male. A pair locked horns and shoved until their limbs twitched and quivered. Eventually the challenger backed away and the owner immediately chased him out of his territory. The neighboring male avidly took up the chase as the loser entered his territory. A third territorial male took up the chase when his turn came, and so on. The unsuccessful challenger ended up running a gauntlet at full tilt across several territories before reaching the ignoble safety of the periphery. Most challengers lost.

Sociobiologically the lek behavior of Uganda kobs is the quintessential stylized territorial system. Females donate all parental investment to their offspring, thus freeing males to spend their extra time competing for opportunities to mate, but females mate *only* with males who hold territory. Originally each male's territory may have been large enough to live on and, in and of itself, a worthwhile piece of real estate containing plenty of forage. Somehow the quality and function of the territory itself became subverted to the male's ability to defeat other males in territorial battles. I imagine that as females became more and more fixated on male prowess and less and less interested in the quality of the ground they held, this very unusual lek system evolved. In this scenario female choice was the major agent of sexual selection.

One predictable by-product of the kob's lek that interests predators is a weakened or preoccupied antelope (hence our stopping here to scan), although any kob who spots a big cat usually whistles a loud alarm. Predators usually are spotted during daylight, but after dark some make kills. Against human hunters, however, kob on a lek are vulnerable, and in the back of my mind I worried about this. (During liberation in 1979, Tanzanian troops all but destroyed the lek system here with their Chinese AK-47's. By 1981, though, a few became reestablished.)

After leaving the lek we saw a vulture-covered mound near a thicket. As I drove up, the carrion eaters reluctantly flapped away to reveal a buffalo cow. Only a few deep scratches marked her flanks, and in her throat were barely discernible puncture wounds inflicted by four canines belonging to a large carnivore. Why had she not been eaten?

Ten feet beyond her I found the head and hooves of a newborn calf; the rest of it was missing. The cow's perineum showed evidence of recent birth. It did not require a Sherlock Holmes to deduce that one or two lions had attacked her primarily to get at the calf. The choice of eating newborn prey was wise because of its lack of parasites. After satiating their hunger, they probably moved to a thicket to lie up for the day. Possibly they would return for the cow. Until her hide was torn open the army of vultures would not be able to touch the meat.

All that meat was tempting; neither of us had eaten meat in weeks. We looked at one another. Karl smiled, "Why not?"

Macabre marabou storks flanked by dozens of vultures waited in a patient arc behind us as we rolled the cow over to expose a haunch not buried in vulture dung. I drew my Buck knife, pulled back a flap of hide, and sliced off nearly ten pounds of meat. I climbed back into the vehicle and slammed the door. Two bushy-maned lions emerged from the thicket sixty-five yards distant and trotted away from us. I suddenly felt foolish, then lucky because they fled, then glad again because we had meat. As I mentally thanked the great cats, the flock of carrion birds raced back to the carcass.

After sundown at Ishasha Camp we wolfed down our boiled buffalo (an immeasurable improvement over the rice, onions, and maize porridge Karl reckoned were adequate to sustain human life) in a hurry so we could investigate lions who had just begun roaring nearby. Only a quarter mile from camp we found two adult males and four lionesses devouring a yearling buffalo. Actually the two males had a monopoly on consumption. They growled, snarled, roared, grimaced, hogged, feinted threats of violence, cuffed, and chased the females away from the carcass to sit and gnaw on a lower leg if they were lucky, or merely to sit and wait if they were not. Apportionment of the spoils was strictly a matter of physical dominance: there was no cooperation and little tolerance.

Their appetites were equally prodigious. What started as 300 pounds of buffalo was reduced to large scraps three hours later, when another pair of younger males arrived and eyed the kill. The newcomers waited with the lionesses, while the older males gorged to repletion. Meanwhile, from beyond the Ishasha River in Zaire, thunderous roars vibrated across the dark savanna.

At midnight two more lionesses, followed by a single tiny cub, joined the group. Now eleven members of the Southern Pride were present at a kill monopolized by two males. Within six minutes these newcomers moved on, but they returned a half hour later. I wondered about the future of that cub. Judith Rudnai found that lionesses in Nairobi National Park, whose litters were decimated to one survivor, generally abandoned or refused to nurse it. By doing so, they quickly cycled back into estrus and became pregnant with a new litter. The sociobiology of abandonment indicates that such females will raise more cubs during their lives than will more sentimental mothers. In addition, the prognosis for solitary male cubs who survive is much poorer than for those with male littermates for other sociobiological reasons (discussed below).

About four hours after we arrived, the old males stood with bloodied faces and manes, abandoned the kill, and slowly stalked into the darkness, their bulging bellies swaying at each step. The two younger males rushed for the remains before the older pair was out of sight. The patient lionesses moved in to search for scraps. Soon the only evidence remaining to suggest a kill were four legs gnawed to the bone, a well-chewed skull, and bits of bone and hide.

Now it was early morning. The two young males joined the three remaining lionesses in their brief and desultory wanderings a few hundred yards from the kill site. We had flashed the light periodically at least every fifteen minutes for the past five hours, and the failing battery of Karl's Land Rover was now little more than a ghost. The subsequent demands on the starter, used every time we followed the lions, killed it.

We sat under a starless sky and fought to stay awake. We were determined to stick with these great cats until the sun drove off the last possibility of another hunt. The chill air crept through my shirt. I strained my eyes to pierce the darkness and monitor the vaguely leonine shapes so we would not lose them through inattentiveness.

"Karl?" It was time for his next fifteen-minute observation.

No answer.

"KARL!"

"Yeah?"

"Are you awake?"

"Yeah."

"It's time to check them again."

"Oh . . . right."

Territorial lions live within the social structure of a pride, which is defined as all the lions who hunt, share kills, mate, and lie up together within the same territory. All members of a large pride (twenty or more lions) rarely collect in one place. And the reason is ecology. The social options of these large carnivores are limited by the size of their prey, which in turn is limited by the size and number of the lions themselves. Adult elephants, hippos, rhinos, and bull buffaloes are rare prey because they are so dangerous to attack. If the entire pride traveled, hunted, and fed together on the smaller prey they do take, most members would be excluded from feeding—unless several prey were killed simultaneously.

The number of lions hunting together at any time is a dynamic consequence of the fluctuations of available prey. In a surprising parallel to party sizes of chimps in Ngogo, Karl's prides hunted in parties averaging three lions. Larger groups were more successful in killing larger prey and in robbing hyenas of *their* prey, but when neither happened, some lions went hungry. Larger parties were themselves robbed less frequently *by* hyenas than small parties, which was another factor pressuring lions to hunt socially. Despite the way in which prey limits how many lions can hunt and feed together, the larger fusion-fission social structure of the pride persists. Apparently it does so due to almost exactly the same set of sociobiological circumstances responsible for the ascendance of the chimpanzee community.

George Schaller (1972), Brian Bertram (1973, 1976), Jeanette Handby with David Bygott, and Anne Pusey with Craig Packer collectively spent about seventeen years observing lions in Serengeti National Park in Tanzania; their combined observations provide an unparalleled portrait. Serengeti lions were either nomadic or territorial; the latter were far more successful reproductively, and it is probably a safe guess that all lions "wish" to be territorial if they can. Like most social large mammals, lion societies retained their females. A female cub in a territorial pride matured to become part of it if she was lucky, but if the size of the pride was already too large for the carrying capacity of the territory, young females were driven out to become nomadic. The reproductive prognosis of a nomadic female is similar to that of females born outside a territory—bad.

Before they attain full adulthood, males who matured in a territory were driven from it by the resident males of the pride. Such young males remained nomadic until strong enough to usurp the resident males of another pride. Often the females of a pride cycled into estrus, mated, and gave birth in near synchrony, which was an important advantage to each surviving male cub because, when ousted, each retained the companionship of his male cohorts, became nomadic with them, and ultimately depended on their assistance and solidarity in usurping males of another pride. This *kinship* among male lions seems the critical factor allowing sexual selection to favor their lethal territorial society because of the advantages accrued through increased inclusive fitness. It is unlike that of the other more than 200 species of carnivores, except cheetahs and wolves, although it is uncannily similar to that of chimpanzees.

Cheetahs and wolves both reinforce this lesson in natural history. Serengeti cheetahs studied by Lori and George Frame exhibited an incomplete version of lion society. Female cheetahs were generally solitary, although sisters, like lionesses, sometimes shared a territory. Male littermates dispersed from their mother's territory to eventually challenge the proprietorship of a territory held by other littermates or coalitions of males. Challengers who lost were sometimes killed. Here again cooperation and solidarity in lethal territoriality were tied to males who were *kin*. But despite ties among *each* sex, cheetahs did not live in a fusion-fission society; no pride structure existed to include *both* sexes.

Wolves, however, did form fusion-fission packs, were zealously territorial (although their territories often were not inherited but established anew by ambitious lone wolves with a single mate), and clashes between neighboring wolves were often lethal. As revealed by the careful research on Isle Royale and in Minnesota by L. David Mech (1970, 1977) and his colleagues, territorial defense and aggression was cooperative among wolves who were either mated or closely related. Although kin relatedness, with its potential for inclusive fitness, seems to be the one essential factor in the *initial* evolution of cooperative territoriality, once it *has* evolved, some social carnivores have improvised on the system to make the best of the hand fate dealt them.

Among Serengeti lions, success in usurping or defending a pride depended on the number of lions in each opposing party. Resident pride males who had lost one or two fellow males, or young males ousted from their natal prides with few, or no, other males were in trouble reproductively. The former males were driven from their prides and some killed. And the latter males on their own had no chance of becoming pride males *unless* they formed alliances. Serengeti lions in this situation often did form alliances with unrelated males, alliances which functioned identically to natal cohorts. Anne Pusey and Craig Packer reported 44 percent of pride males consisted of such coalitions. A kin cohort or coalition was vital for reproductive success. This mutual dependence among males explains why they so often are seen together. Interestingly, captive male chimpanzees studied by

Frans de Waal exhibited similar, but protean dominance coalitions which were pure Machiavellian opportunism and formed solely to gain more opportunities to mate.

The lion pride is clearly a female affair: a permanent matrilineal proprietorship of a hunting territory which fortunate lionesses never leave. Females contribute nearly all the parental investment in their cubs: they conceal them, nurse them (many lactating lionesses of a pride suckle all its cubs indiscriminately), defend them, and train them to hunt, thus becoming ecologically independent of males (who, after all, are serious competitors at kills) and freeing males to devote their lives to competing reproductively against other males. This inequality of parental investment in the sociobiological equation has allowed sexual selection to favor large size among males, plus their unusual strategy of remaining lifetime companions with their littermates, which, because of inclusive fitness, are reproductive allies.

But adult males, despite only temporary residence in the territory, are important to the females as territorial defenders and fathers of their cubs. Because successfully invading lions not only kill resident males, but may also kill the pride's young cubs (as do cheetahs and wolves) in order to recycle the lionesses into estrus, it is in each female's best reproductive interests to retain the same males for as long as possible. Pusey and Packer reported that the tenures of Serengeti males averaged only two years, although six years was maximum. As was the key to territorial success among Tanzanian chimpanzees, larger cohorts or coalitions of lions retained tenure longer and were the most successful reproductively. Hence the amazing tolerance shown by lionesses toward males who so often chased them from their kills to take the lion's share. This picture came from the Serengeti, but it also described the lions sleeping in front of Karl's Land Rover and those studied elsewhere except in the vast Kalahari Desert. Mark and Delia Owens reported that, while during normal seasons pride behavior resembled that at the Serengeti, during the worst of the dry seasons Kalahari prides sometimes split up and temporarily remixed with members of other prides and even occasionally mated with their males.

Lions are the *only* known fusion-fission society of nonhuman mammals other than chimpanzees in which all adult members of both sexes breed *and* in which kin-related males form cooperative breeding coalitions based on territorial defense to the death. The main difference, and a very important one, is that, unlike male lions, male chimpanzees (at least in Tanzania) *inherit* their territory, are dependent on it utterly, and their reproductive fortunes wax or wane directly with their ability to hold or expand it. Their territory is absolutely vital. It is small wonder why males would kill for it. Were the nonprovisioned chimps in Kibale territorial, and to the same extent? While it seemed to me that they had to be, I still did not know.

The battery was dead. One of us had to go out there with the lions and crank the engine by hand to start it and recharge the battery. I knew those five lions were still hungry: I could almost count their ribs.

"They probably won't attack you, Karl. You're too skinny."

"Thanks. . . ."

I stood outside the driver's door with one foot on the gas pedal and trained a very feeble beam from the spotlight on the resting lions while Karl cranked. After several arm-wrenching failures the engine finally burst into mechanochemical life. As usual, the lions ignored us.

For the remainder of the night they napped in one open area after another. They strayed only a quarter mile from the site of the buffalo kill and made no serious attempt to hunt again. Several times hippos plodded within forty yards of them, but only once did a lion tentatively follow a young one. He gave up quickly, possibly due to the presence of so many adults.

After a subjective eternity of studying lion napping in the dark, the sun shot a golden corona above the horizon north of the Virunga Volcanoes straddling the borders between Zaire, Uganda, and Rwanda. When sunrays hit the lions, they rose, stretched, then slowly padded to a nearby papyrus swamp to lie up for the day. After twelve hours with them, we turned back toward our hut, also to lie up for the day. My forty-four hours of observing lions during the last week had again renewed my perspective on the apes of Ngogo and made finding the missing pieces to the puzzle of their social system even more desirable. These thoughts would hold me during the long drive ahead.

— 14 —

Return of the Killer Apes

April brought the most rain ever to Ngogo, eleven inches, but this morning was clear. The forest's sun-dappled greens were shot with the flights of gaudy birds and butterflies. I walked south on trail C.5 and wondered if the figs on the gigantic *dawei* ahead were now ripe. Nearly a year had passed since its last crop had ripened; it had to repeat itself soon. It was the second largest fig tree I knew of in the region. When its tens of thousands of figs did ripen it would become the busiest place at Ngogo. For the chimpanzees it would be an irresistible magnet.

Because I really did not expect the figs to be ripe already, I was less stealthy than I should have been near the junction of C.5 and 8.5. Three chimpanzees suddenly scrambled out of the *dawei* and melted into the forest. I focused my field glasses to identify a fourth ape, who continued to forage. Gray turned her wrinkled and scarred face toward me, blinked calmly, then plucked another fig. Apparently she finally had forgiven me for the lightning and thunder exploding near her seven months earlier. I remained north of the *dawei* on a slope leading down to a sandy-bedded little creek marking the edge of a swamp sixty-five yards wide. On the southern edge of this swamp, along trail 8.5, was a much better vantage point. During my next eleven days there the apes of Ngogo would tutor me in a review course in chimpanzee social dynamics.

Gray foraged unhurriedly and methodically, picking only the ripest figs from each terminal cluster. After twenty minutes she paused, stared down toward the south, then pant-grunted spiritedly and shrieked for half a minute. R.P. and Phantom, a large juvenile male, climbed into the *dawei* and unceremoniously examined fruit not far from her. A moment later Gray lip-smacked while approaching R.P., settled next to him, and groomed the back of his shoulder. Despite probably being hungry, R.P. delayed his foraging for half an hour to groom with Gray. As usual, Gray used her left hand primarily rather than her partly crippled right hand. Just as usual, both while Gray groomed him and while he reciprocated, R.P.'s left hand crept around his side to pat at the edges of his wound.

The repeated closeness between this young headstrong male and this ragged old female prompted me to wonder if she was his mother. I had no real evidence. But their mutual attraction was not a sexual one. Gray never cycled into estrus; her reproductive days were buried in the past. The concept of her being his mother appealed to me. Both apes had accepted me early, but independently of one another. I tried to imagine Gray as a svelte young mother carrying a cute little R.P. It was difficult. R.P. was not cute and Gray was ancient. Whether or not Gray was R.P.'s mother, she probably had known him since he was an infant.

As they groomed, an adult male redtail with a short tail rushed through the crown from the east, where a harem of redtails was foraging. Without a pause he hastened to the west. Seconds behind him another redtail male pursued intently, this one with a normal tail. Most likely an attempt had been made by one male to usurp the other as harem master. But I did not know the harem male well enough to tell whether the attempt had been successful. As mentioned in Chapter 2, Tom Struhsaker had observed infanticide by newly ensconced usurpers of redtail harems at Ngogo, and Tom Butynski (1982b) would later make similar observations among blue monkeys here. If a takeover had occurred, the new harem master might soon kill the youngest infants in it to accelerate their mothers' sexual cycles into estrus quickly, and, as a by-product, reduce the competition faced by his own future infants should he reign long enough to sire any. Minutes later the redtail with the normal tail returned through the *dawei* to rejoin the harem. As he passed, Gray broke off grooming and resumed foraging.

Males on the run were more common at Ngogo than in the rest of the forest. Only one group of blue monkeys inhabited Ngogo, which seemed the southern boundary of their range in Kibale Forest. But fifteen to twenty redtail harems ranged within the trail-grid system. Male blue monkeys pushed southward by the more numerous, aggressive harem male blues in the north found themselves in an awkward predicament. To reproduce they had to usurp a harem male. But harems of blues were virtually non-existent at Ngogo. The young blue bravos took their next best opportunity by challenging a redtail harem master, the next most closely related species. Because blues outweigh redtails by nearly 50 percent, they often won and mated with a few of the redtail females, which seems strange because the two look as different as night and day and redtails are generally afraid of the larger blues. Their hybrid offspring look more like blues, but they were integral members of their natal groups. These hybrids were relatively common at Ngogo, but very rare elsewhere.

Two hybrids frequented the vicinity of this *dawei*. One was a female member of the redtail harem from which the short-tailed male had just been chased; the other was a recently matured male who alternated solitary spells with brief associations with a redtail harem which ranged further south. I was uncertain as to whether he was attempting to take over the harem or whether he had been born in it and merely was lingering. My previous sightings of him a year earlier made me suspect now that he was lingering. Either way, the redtail harem master repeatedly chased him off to live the vulnerable life of a bachelor.

After an hour of consuming figs, Phantom climbed to R.P. and started grooming his back. This session lasted for another thirty minutes. Afterward the males foraged again. Two hours after they had arrived, R.P. and Phantom descended the *dawei*. Gray watched them disappear into the forest, then she carefully selected more figs for another half hour, by which time groups of mangabeys and redtails had filled the tree like a rising tide. Within minutes of their arrival Gray climbed down and was lost to view. She had consumed scores of figs during the past three hours.

This *dawei* was in its finest hour, the single week of the year when no place in Ngogo compared to it for sheer quantity of top-quality calories. The local redtails and mangabeys would visit it almost every day. Even so, chimpanzees would outnumber them in hours spent in the tree. Late that afternoon, for example, Blondie, Butch, and Bess, plus Kella and Kirk arrived to forage. Bess spotted me on her way up, and I felt a sinking sensation in anticipation of her doing an about-face to flee. But she merely paused for several seconds, then continued into the tree. And, when her mother noticed me a half hour later, the lure of this fig tree was so great that, instead of panicking as she usually did, she hastened to retrieve Butch but then continued to pick figs and only glance at me occasionally.

As rain poured from a slate sky, R.P., Phantom, and three other chimpanzees whom I could not identify because of the rain joined them. Most of the apes remained to forage until premature dusk caused by heavy rain drove several of them to seek nesting spots in a large *Monodora*. Its big-leafed crown appeared to be in fits. Above the pounding of the rain and the even louder electric stridulation of cicadas, branches snapped, boughs waved and vibrated, and limbs sagged as three chimpanzees simultaneously constructed nests for the coming darkness. They were destined to spend an uncomfortable night. I praised human technology as I hiked toward my mud hut.

The next day the traffic at the *dawei* resembled a jungle version of a white sale. The tree received twenty-six visits by chimpanzees and more than fifty by monkeys. But that was only a beginning. On Day Number Three, R.P., La with Lysa, S.B., Gray, Zira, Stump, Blondie with Bess and Butch, Farkle with Fanny, Fern, and Felony, Eskimo, Shemp, Newman, Ashly, Mom with Munch and Mac, Spots, Owl, Ita, plus a half dozen other chimpanzees arrived at the *dawei* for a total of forty-four visits. During the eleven days and 123 hours that I monitored visits to it, the *dawei* received 212 visits by chimpanzees, 278 by monkeys, and untold visits by squirrels, fruit-eating birds, and hammer-headed bats.

Over this period, thirty-four of the forty-six chimpanzees I had named, plus another ten whom I did not know well, visited this tree. This was the record for all my vigil trees. These visits confirmed my analysis of the relationship between the size of a food patch and the intensity of its use by the apes: larger fruit trees attracted larger aggregations, longer stays by each chimpanzee, and more return visits by each ape. And this *dawei* emphasized an important lesson: the largest food trees were much more than bigger food patches, they were important centers, almost catalysts of socialization for the apes.

The basic social structure of chimpanzees has been one of their more recondite aspects, one that has eluded or confused many of the field workers who sought to clarify it. Similarly, even after long days at Ngogo, my data often gave no hint of significance on their own. I frequently looked forward to my full analysis, the chore that would create order from apparent chaos and justify my research. This gathering of the Ngogo Community gave me a good look at the light at the end of the tunnel.

The parties of chimpanzees visiting and departing this *dawei* followed the same nonrandom pattern they had been following all along. But here it was even more blatant. The apes chose partners based on their sex. Males clearly preferred other males as companions; females preferred other females, and among them mothers preferred other mothers. But not only were partners of males and females biased sexually, even their *patterns* of visitation to the *dawei* differed. On average, males visited on fewer days and fewer times per individual than females. But on those days when they *did* visit, they revisited the *dawei* more often than did females. In other words, each male apparently was not in the neighborhood of the *dawei* as often as each female but, when he was, he made *more* use of it.

These sexual differences were in concert with my overall analysis of activity patterns (see Chapter 4). Compared to females, males foraged during more of their total time, traveled more, and rested less, a full hour less daily. Although statistically significant, these differences were not so dramatic that a female traveling with males would be left behind in a panting heap on the path. But a mother carrying her infant could count on a more difficult and metabolically expensive day with males. These his-and-hers differences in companions and activity profiles were solid evidence that males and females actually had different life-styles,

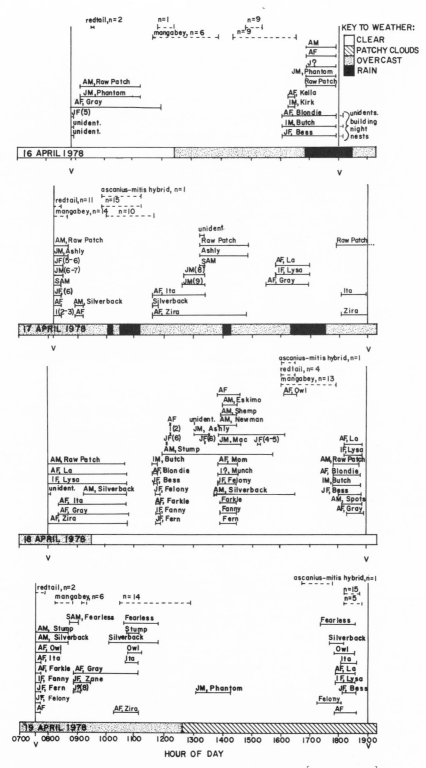

HOUR OF DAY

[CONTINUED]

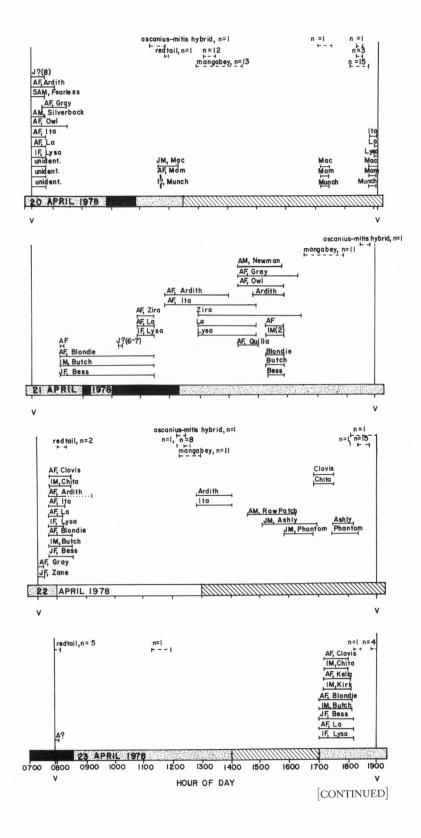

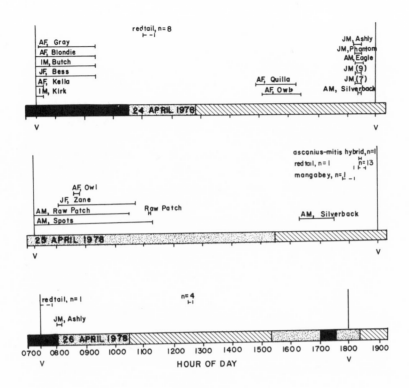

Eleven-day record of visits to a fruiting *Ficus dawei* by Ngogo chimpanzees and monkeys. The vertical lines labeled V designate the period of the author's vigil per day. Horizontal dotted lines indicate that the chimpanzee was resting near the vigil tree. Dashed lines indicate the presence of monkeys in the vigil tree.

based on different priorities. Why should the apes of Ngogo have been so choosy about their companions?

Because being social exacts a cost. The basic arithmetic of resources is this: if a chimpanzee elects to travel with another, it can hope for no more than half the food they find, unless the food is unusually defensible (a captured monkey, for instance) *and* one of the apes is dominant over the other ("dominance" is used here as defined by Edward O. Wilson in *Sociobiology, the New Synthesis*: "To dominate is to possess priority of access to the necessities of life and reproduction"). Dominance interactions among the apes of Ngogo were surprisingly few, which, I suppose, is not surprising after all *if* most of the chimpanzees

243

were either satisfied with their rank or convinced that they should bide their time before attempting to upgrade it. Likewise, if a chimpanzee travels with nine others, it can hope for no more than a tenth of the food they find. During seasons when food patches are small, each additional ape requires its party to travel a bit further than before to visit enough patches to feed everyone. The calories expended for this extra travel also escalate the metabolic requirements of each chimp, requiring them to travel even further, and so on. As discussed in Chapter 4 and expanded in Chapter 9, unless huge food patches exist, such as fruiting *mucuso* or *dawei* trees, chimpanzees often cannot afford the cost of traveling in large parties.

I reckoned the Ngogo Community contained at least fifty-five to sixty members. If all of them were to travel together, it would be impossible to find enough fruit to feed them all *and* have enough daylight left to travel the required distance connecting the food sources, no matter how efficient the itinerary. Large troops of baboons and red colobus manage to travel as a unit only because they eat very common foods, such as grass corms (on the savanna) and leaves. By contrast, 78 percent of the diet of the Ngogo chimps was fruit, which is rare and ephemeral. And, as fruit decreased, so did the size of traveling parties. By traveling in smaller parties or alone (which happened 44 percent of the time) during lean seasons, the apes spread over a wider area and minimized competition among themselves for each fruit tree, just as they spread out when they fed together in the *same* fruit tree (Chapter 9). *Without this critical adaptation of fusion-fission sociality the chimpanzee community could not exist.* The bottom line is this: because companions are expensive, each chimpanzee should choose to travel with other apes who benefit it the most. Which ape benefits another the most?

A chimpanzee mother normally gains little or nothing from traveling with an adult male (as discussed in Chapter 12), even if he is the father of her infant. She may even be adversely affected by having to compete with him for food. This independence of the sexes has freed each to pursue divergent life-styles. Because the metabolic needs of an adult female carrying and nursing an infant (or pregnant) equal or even exceed those of an adult male who travels the same distance, the greater traveling by males cannot be explained as a result of required additional searching demanded by greater metabolic needs. Nor can it be explained by males at Ngogo having traveled in larger parties than females; parties were nearly equal in size. Each sex traveled on average in parties containing two or three apes (not counting infants). So males were less beneficial traveling partners to females because they traveled too much. Why did they?

The sociobiological explanation for why male chimpanzees prefer one another rests upon them cooperating and relying on one another to exclude other males outside their kin group or coalition from mating with their females, threatening their offspring, or harvesting any of the limited resources in their home range required by the resident males and the other members of their community. The hypothesis I proposed for my research, and in this book, is that they accomplish this exclusion of non-community members through cooperative territoriality. For that purpose alone males must travel extensively, much more than females. Males must patrol boundary regions to dissuade invasion, and, for their mutual protection, they must do so together. But males must travel more than females for another reason entirely, to locate females during their rare periods of estrus. Females need not travel for either one of these reasons, hence their extra hour of resting each day.

But how can we know which reason best explains why males traveled more? More than half the times adult and subadult males went anywhere at Ngogo, they traveled with one or more other males. It is important to remember that, by doing so, they compromised their rare chances to sire offspring. Despite the additional solidarity that sharing in one another's mating success through inclusive fitness may offer, it is obvious that most males preferred to monopolize an estrous female. Instead of being clever loners, however, males in Kibale were sophisticated socialites who exhibited several very unusual adaptations to enhance their solidarity: they preferred one another as traveling companions (even as young as juveniles); males groomed one another longer and more frequently than they groomed with females; males generally tolerated one another's opportunities to mate, almost cooperating sometimes in serial mating; males pant-hooted their long-range calls ten times more frequently than females, thus allowing one another to monitor their whereabouts when separated, and, at times, probably warning away alien males; males were the only sex to give food calls advertising the presence of superabundant fruit patches to other community members; and, when foraging together, males exhibited an absolute minimum of dominance interactions over food. All of these special adaptations promote solidarity in ways unknown among nearly every other species of mammal, but they also compromise each male's ability to monopolize an estrous female. Because males travel *together*, and by doing so compromise their individual mating success, it is difficult not to conclude that the primary reason for their extra travel is to defend their territory, and the primary reason they travel together is for mutual protection from alien males. Almost certainly, however, at times they also travel extensively to avoid missing opportunities to mate with estrous females.

When traveling for territorial maintenance, defense by the chimpanzees need not include physical clashes with alien males. It normally would suffice simply to pant-hoot, build nests, defecate, and otherwise provide signs and evidence advising alien males that this real estate is occupied by numerous vigorous males with solidarity. My guess is that only when the males of one community consistently detect a small showing by the males of a neighboring community do interactions lead to combat (see below). Of course, it is during such times that solidarity between males of a community is indispensable.

The obverse side of the coin of companionship is the active preference that Ngogo females exhibited for each other. In both Gombe and Mahale adult females socialized most with males. In fact, the Ngogo females' preference for one another much more closely resembles that of bonobos (pygmy chimpanzees). But several explanations may explain this preference in Kibale: 1. Because females hold lower status than males, they are likely to be cheated by them during grooming, whereas grooming between females generally is scrupulously fair. 2. Females also tend to lose to males in any contest over defensible foods. 3. Again, a female who elected to travel with males would expend more calories and have less leisure time than with females, a disadvantage for a mother with dependent young. 4. If males occasionally clash territorially, then a female traveling with them would expose herself and her young unnecessarily to danger. 5. And, in the interest of the social development of her growing children, a mother should associate with other mothers so their offspring can play together to learn and develop social identities. This is especially important for male infants who must attain solidarity during adulthood. 6. It is possible that some adult females at Ngogo were related to one another as closely as were the males, either because a young female never emigrated and continued to socialize with her mother (as in the case of Zira and Gray?), or because the females present at Ngogo immigrated to it but were born and raised together in the same neighboring community. Their preference for one another might be due partly to family ties. 7. Undoubtedly, real friendships also play a role in whom a female choses as a companion. I suspect *all* these factors influenced Ngogo females. But part of their choice may have been by default. Because males did not normally choose nonestrous females as companions, the females' choices were narrowed to each other.

The Kibale females' penchant for one another was at least equal to that between males for one another. Chimpanzee society at Ngogo was sexist in the extreme, due to apparent sexual incompatibilities in activity patterns caused by ecology and by extremely divergent reproductive strategies. Comparing the social choices by females in Kibale with those at Gombe and Mahale, however, is complicated because provisioning the chimps with food at the latter two study areas may have removed some of the ecological constraints that keep the sexes separate. I am convinced that Kibale females formed a real society that was mutually beneficial. It is difficult to imagine why females in other habitats would not do likewise.

Although it would be impossible to understand why chimpanzees do what they do unless we looked at sexual differences, especially in social choices, we must look even deeper to appreciate the level on which they base their decisions—to the level of the individual chimpanzee. As in human societies, it seems that every member of the Ngogo Community was either a friend, indifferent, or distasteful individually to every other member.

As day after day chimpanzees returned to feast at the *dawei*, then groom one another while their infants wrestled and dangled precariously, or one large male bluffed another as if required to by some quota, I concentrated on the more subtle exchanges in hopes of understanding relationships between some of the apes.

This, of course, was frustrating because I would never be able to know them for sure. The one that tormented me the most was who the juvenile, Zane, was in relation to Gray. And here at this fig tree they revealed the biggest, but most confusing clue of all.

I had seen Gray traveling with Zira, whom I suspected was her daughter, if not her best friend, twenty-nine times by now. But eight times Gray had traveled with Zane (whom I had named for a suspected relationship to both Gray and Zira). Their association was undoubtedly special. Although I had seen Gray travel with fifteen apes other than Zira and Zane, the only other companions whose frequency could be stretched to "frequent" were Owl, with five times, and Ita and R.P. with scores of only three times each. Zane was number two in Gray's life. Why?

One morning I arrived at the *dawei* and found them together. When they eventually descended, Gray led the way along the horizontal bole of a *natalensis* arching across the swamp, then she leapt from its end to a tall sapling nearby. Zane followed, but when she reached the gap at the end she stopped cold, afraid to leap it. Gray glanced at Zane, climbed back up the sapling, pumped it to swing it to and fro until its arc took her close to Zane, then Gray grabbed the *natalensis*. She gripped both trees under tension and held her spread-eagled pose until Zane had scrambled over her body, using it as a living bridge. Then they both descended the sapling.

In view of how unabashedly self-centered chimpanzees are, except toward relatives, Gray's bridging for Zane should have been revealing. Jane Goodall (1968) reported Gombe mothers bridging occasionally for their infants. Herman Rijksen (1978) reported Sumatran orangutan mothers doing the same thing. Allison Jolly mentioned both spider monkey and baboon mothers assisting their young by bridging. Rijksen even saw female orangutans in estrus bridging for their male consorts! But in all cases the female who bridged had a major interest in the one crossing the bridge.

So . . . I thought, Gray cares about Zane (which was something I knew already). But how much, and why? Was Gray the simian equivalent of the kindly old grandmother who impartially helps youngsters in need? Only tiny pieces of the puzzle were available and they had to be glued together with speculation.

Only two days later the exact situation recurred, but this time Zane was not present and Gray was traveling with Blondie, Butch, and Bess. During all their previous interactions Gray's behavior toward Blondie had convinced me that she was not fond of her at all. In fact, I was very surprised to see them traveling together. With Butch hugging her belly, Blondie led the way along the horizonal *natalensis*. She reached the end, jumped across the gap, and climbed down it.

Gray followed, also leapt the gap, climbed partway down the sapling, then stopped where she had previously to bridge for Zane. Bess brought up the rear, reached the gap, and hesitated. She made a few undecided, false starts to leap across. Then she pumped the *natalensis* up and down in an attempt to use its spring to catapult her across, but she was too afraid to let go and jump. Then she tried climbing out on a side limb to get closer to the sapling, but it drooped too much and she got no closer. Four minutes had passed while poor Bess tried to get across that gap, and all the while Gray sat impassively ten feet away, as if in calculated and studied disinterest. Then Gray descended the sapling and disappeared after Blondie.

Bess redoubled her pumping of the *natalensis* but could not bring herself to make the leap. She tried every approach again that she had tried before but could not manage to actually jump. She was almost frantic but made no outcry. I wanted to climb the sapling myself and help her across, but Bess took her cues from Blondie, who considered me the devil. Bess would flee from me. Finally, eight minutes after Blondie had left, Bess turned and ran back along the *natalensis* to the *dawei* and climbed down the twisted flutes of its hoary bole to vanish in the swamp. I hoped that she would grow up to be a better mother than Blondie.

Gray's studied rebuff of Bess in her moment of need was in total contrast to her maternal solicitiousness for Zane (both juveniles were the same size) only two days earlier. While Gray's rebuff opens questions about the level of mental evolution required to invent deliberate cruelty, the reason for Gray's obvious concern for Zane remained a mystery. Because she did not otherwise behave as Zane's mother, and Zane traveled with her only a quarter of the time, I wondered if she was her grandmother, but even this seemed unlikely. The more I considered it, the more it seemed to me that Zane was the orphaned daughter of a "friend" or even sister of Gray, whom Gray had only incompletely adopted.

Because positive relationships involving apparent altruism between relatives are reinforced genetically by inclusive fitness, to label them as "friendships" is inappropriate. Although their interactions may seem the most friendly, because of genetics, they are enlightened self-interest. Genuine friendship implies a consistently positive relationship between individuals who are not related. In sociobiological terms, in fact, friendship has been analyzed and relabeled *reciprocal altruism* by Robert Trivers (1971), and it describes relationships or partnerships between unrelated individuals in which each member performs favors for the other *with the expectation* that those favors will be reciprocated when appropriate. The process of reciprocal altruism explains the sociobiological basis for cooperation and solidarity within coalitions, for instance, of unrelated male lions who defend a territory, or of unrelated female chimpanzees who groom and raise their infants together.

My two years with the apes of Ngogo were too little time to rule out the possibility of any two chimpanzees being related. Even twenty years of observation at Mahale and Gombe were too few to rule this out among some of the older adults. So, because of the real possibility that "friendly" chimpanzees at Ngogo could have been cousins or half siblings, I hesitated somewhat to think of their relationships as true friendships, even though the genuine article, especially among females, almost certainly applied.

Still, many chimpanzees consistently groomed or traveled with one or two preferred companions: Blondie with Ardith, Gray with Zira and Zane, R.P. with Stump, and Stump with S.B. and R.P. When I analyzed these associations between individuals I found some unsurprising (by now) tendencies: social bonds between chimpanzees of the same sex were reinforced nearly twice as often as bonds between chimpanzees of the opposite sex. In view of all the other consistently sexist patterns, anything else would have been surprising. If males never emigrated from their community, the bonds between them are easily explained as being reinforced by inclusive fitness. Partnerships among females were more difficult to explain, except through reciprocal altruism. But I rarely thought of most of these partnerships as sociobiological statistics; to me they looked like friendships. But is it possible that chimpanzees related as distantly as cousins recognize one another somehow, even unconsciously, and form genetically biased "friendships"?

Such a possibility has been suspected only recently in the animal world, but already the answer looks like yes. Andrew Blaustein and Richard O'Hara found tadpoles of the Cascades frog could distinguish siblings from nonsiblings, even if they had never seen them before. Gary Beauchamp, Kunio Yamazaki, and Edward Boyse found not only that mice could discriminate their close kin by smell and subsequently preferred more distantly related mice as mates, they also identified the genetic material of the major histocompatibility complex responsible for kin recognition. Admittedly, tadpoles and mice are a long way from chimpanzees. What about primates?

A team led by W. M. H. Wu demonstrated that infant crab-eating macaques could recognize paternal (i.e., half) siblings they had never seen before and preferred to sit with them. Similar investigations of chimpanzees and most other species of primates have not been done, but it would not be surprising if wild apes did possess some degree of innate ability to help them recognize kin. Obviously they do possess the ability to recognize behavioral relationships between the apes around them, for instance the consistent way in which an ape's mother interacts with that ape's maternal siblings. But what about paternal siblings, as among the macaques, or cousins? Perhaps they are able to recognize biochemical similarities, but even if they are not, males cannot help but recognize most of the other chimpanzees of their natal community as having grown up in it or not. This simple identification, though it need not be cognitive, is critical to the male kin group identity which is the backbone of the community. And this community appears to be unique among nonhuman primates.

Chimpanzees are only one of four species of the nonhuman Hominoidea (the taxonomic superfamily containing the great apes and man). Bonobos, also called pygmy chimpanzees (though they are equal in size to common chimpanzees in Central Africa), are very closely related, having split from a common lineage perhaps within the last two million years. Bonobos have not been studied to the same extent as chimpanzees (I do not wish to call them "common" chimpanzees because they *are* no longer common), but a recent comprehensive symposium volume edited by Randall Susman provides a fair basis for comparison. In short, bonobos exhibit nearly an identical ecology, plus a community social structure apparently similar to that of chimpanzees, but with the following differences: relationships between females are somewhat stronger than among chimpanzees (although appearing similar to Ngogo females' relationships); relationships between males and females seem much stronger and more durable than among chimps; parties of one or both sexes last far longer than among chimpanzees, sometimes for weeks; all-male parties are extremely rare (instead of the norm as among chimps), even though bonobo communities appear unequivocally territorial. It is uncertain whether territorial defense includes potentially lethal combat. In many ways, especially in the closer relations between male and female adults, bonobos seem more human-like than do chimps. In other ways, territorial defense, for instance, they appear less so. More research is needed.

Gorillas are also closely related to chimpanzees, although, based on DNA-DNA hybridization, their split from the common lineage of the Hominoidea occurred possibly at least eight million years ago, earlier than the line which led to the human-chimp-bonobo lineage—and the common ancestor leading to man. Socially and ecologically gorillas are quite divergent. The primary foods of mountain gorillas consist mostly of common herbage which is relatively easy to find and available in such massive quantities that a group of a dozen gorillas normally has no trouble finding enough food for everyone within a daily itinerary of only a third of a mile. Very little ecological pressure works against group cohesion, and, possibly as a consequence, gorilla groups are very stable and tight socially.

Reports from George Schaller, Dian Fossey, and Sandy Harcourt (1978, 1979) characterize mountain gorilla society as one-male harem type, though occasionally a second or third male, usually sons or siblings of the harem male (members of a male *kin* group), may also be present. Quite unlike normal primate harems composed of kin groups of females which more properly would be termed matrilineal social groups, mountain gorilla harems are true harems. They are created through the ambitions of the harem male, who normally begins his career by kidnapping, herding, or seducing females, both adolescent and adult, from established harems. Females recruited generally fit into one of two categories: young females ready to emigrate from the natal harem to mate with a male unrelated to them, and low-ranking adult females whose infants may be threatened by higher ranking females or coalitions of higher ranking females, who may be infanticidal. Although harem males are not territorial, they generally lead their families on routes intended to avoid other harems. Of course harem males are extremely intolerant of other, non-related males, especially those without harems of their own, who not only present the risk of losing their females through abduction, but present an equal threat of killing the harem's young infants during a rampaging intrusion and attempted abduction of females. Because of the males' reproductive strategy of drawing females to them, plus the tolerance exhibited by some males for their mature male kin and their draconian penchant for killing the infants of unrelated males, mountain gorilla society resembles a very incipient, undeveloped form of the chimpanzee community.

By contrast, orangutan society deviates completely from those of all the other great apes. Despite an ecological relationship almost identical to that of the fruit-dominated chimpanzees and bonobos, multiple reports by Peter Rodman, John MacKinnon, Herman Rijksen, and Berute Galdikas indicate that both Bornean and Sumatran orangutans are basically solitary. Galdikas presented the most long-term data on females, which indicated that, rather than being strictly solitary, female orangutans spent about 6 percent of their time associating with one another. Because, unlike all other male great apes, male orangutans emigrate from their natal areas whereas females tend to remain, the females Galdikas reported most likely were kin. Thus orangutan society most closely resembles the most common system among nonhuman primates, and because, as a best guess, the split of the orangutan lineage from that leading to the African apes and man occurred approximately fifteen million years ago, their social system may be representative of the ancestral, "primitive" primate system also typical of the earliest Hominoidea and other nonhuman primates in general.

Looking at ecology alone, we would expect orangutans and

chimpanzees to be almost identical socially. Because females provide virtually all parental investment, the reproductive strategy of adult male orangutans consists of one of two options: to remain in the home ranges of a couple of females and attempt to exclude all other adult males by warning them with long calls and by besting serious challengers in combat, or to wander constantly over the home ranges of many females to increase their odds of finding one in estrus—and impregnating her at the expense of the resident male. Some females apparently are attracted to the long calls of "their" resident male and may seek him during estrus. Because orangutans are far more adapted to arboreal life—with permanently curled fingers and toes which impede rapid or long-distance travel on the ground—they usually travel in the trees and cover less than a fifth the distance that chimpanzees do daily. Except for rare temporary associations between adult females, their limited mobility appears to lock their society forever in the fission state, because in most ecological situations they simply cannot travel far enough together to visit enough fruit trees to feed a group of them. And as is apparent from the reproductive strategies employed by males, socialization with other males would be counterproductive.

This overview of the social systems of the living great apes brings the salient aspects of chimpanzee social life into sharper focus. Throughout this book I have discussed social behavior in Darwinian terms, as if individuals were preoccupied with the reproductive consequences of their acts—or at least *acted* as if they were. I have repeatedly referred to social mores sociobiologically as reproductive strategies. This is no accident. The great geneticist Theodosius Dobzhansky once wrote that nothing in biology makes sense except in the light of evolution. And most biologists agree with him. And without a doubt chimpanzees do have reproductive strategies. Male chimpanzees especially reveal the lengths to which evolution can take such strategies. As I mentioned in Chapter 13, the males of a community are almost certainly a kin group. Even the males in Kibale exhibited solidarity among themselves that otherwise defies explanation. While males of almost every other species of mammal lose their male kin upon maturity and enter the reproductive contest alone, male chimpanzees retain every male family member and constantly cultivate their solidarity. Through inclusive fitness they breed as a *communal* unit. Albeit males still prefer to monopolize females, they frequently subjugate this priority to retain the solidarity vital for them to hold a *hereditary* territory containing their mates and all the resources necessary to raise their off-spring. No lone male could hope to compete long in this system, let alone survive. The data from Gombe, Mahale, other study areas, and now my data from Kibale reveal that male chimpanzees have taken the common reproductive strategy of male territoriality a quantum leap in social evolution beyond the rest of the animal world.

Even so, argument persists about the exact nature of the chimpanzee community. Richard Wrangham (1979a, b) analyzed the ranging patterns of males and females at Gombe and found that males traveled over a significantly greater area than individual females, who ranged as if the males' territory had little meaning to them. Harold Bauer reported that adult males at Gombe were more gregarious than females, becoming considerably more so with advancing age. Wrangham and Barbara Smuts noted the Gombe females spent most their time alone, whereas males spent most of their time with males. Wrangham concluded that the chimpanzee community was strictly a male phenomenon and, from a male perspective, females in the system were simply reproductive resources. It is easy to see the logic of this males-only model, but it fails to consider several aspects of female sociality and ignores several complex social realities faced by females.

Because the targets of aggression by territorial males have included the infants of strange females, adult females must be aware of community dynamics and community boundaries to avoid losing their infants. Wolf and Schulman reported that Tanzanian males universally recruited young females, but repelled, attacked, and even killed old females entering their territory. Clearly it is in a female's best reproductive interests to emigrate from her natal community early in her reproductive career. Males in Mahale attacked the infants of some *peripheral* females, *even if* they had mated with them during their period of conception. This emphasizes how important it is for females to seek home ranges in the *core* of the males' territory. This may be difficult, however, because established females in Gombe sometimes formed alliances with one another and cooperated in repelling new immigrant females from settling in a saturated core. This probably also explains in part the preference of Ngogo females for one another as traveling and grooming partners. The formula for a successful reproductive strategy for female chimpanzees might read: immigrate as a young adult, settle only in a core range, be prepared to resist female aggression, and establish alliances with resident males *and* females as soon as possible. At the same time it is important for new immigrants to assess the quality of the food resources in the new community and note its number of males to gauge its future viability as a social entity. All of these concerns of females demand not only that they recognize the community nature of things, but that they carefully manipulate their own situation by becoming part of the community dynamics.

The concept of parental investment is indispensable for evaluating the differences between what males and females do in a society. As discussed in Chapter 12, the inequity in parental investment between the sexes among chimpanzees seems wildly disparate. Males donate their brief courtship, mating, and sperm, while females donate nine months of pregnancy, four or more years of infant care, plus additional years of association with their juvenile young. Chimpanzee society is ripe for activists seeking female rights.

But by maintaining a territory, males also ensure for the community's females a safe area in which to forage and raise their offspring. And by taking older juvenile males *with* them, an extremely common pattern in Kibale and in all other long studies, they train the females' sons (and their own) in becoming successful males by furthering their ecological educations and tutoring them in territorial realities. Adult males have a vested interest in these immature males becoming successful graduates because so much of their reproductive success hinges on the future of their communal territory. Because the strategy and form of parental investment by males differ so much from that of females, some of it is too camouflaged to be recognized as parental investment. Even so, males contribute only a small fraction of the parental investment their offspring receive, and use their freedom to the hilt in competing territorially with other kin groups of males. Hence the superficial appearance that males are the only active community coalition, while females are merely separated and independent reproductive tools of the males. But the biological realities faced by females dictate that they also be integral members of the community. In my view and those of Jane Goodall (1986), Toshisada Nishida and his colleagues (1985), and others, the functioning community as a social unit is composed of both sexes, who contribute to its existence and benefit from it in different ways, but depend on it equally for reproductive success.

For the community system to succeed males must have good mobility, intelligence, long-range communication, food calling, tolerance of one another's opportunities to mate, positive affiliations increasing solidarity (such as mutual grooming), sophisticated behaviors for individuals to split from and regroup in traveling parties, and cooperation in territoriality. So far I had seen all of these in Kibale Forest except the last crucial element, active territorial defense. Perhaps because Ngogo was surrounded in the north and east by grassland which chimpanzees rarely visited, it was an inappropriate part of the community's home range for territorial clashes to occur. Not having seen active territoriality does leave the question in Kibale open, but no other explanation accounts for the striking differences I observed in activity patterns between males and females, the cooperation and tight solidarity among males, and all the other sophisticated adaptations and behaviors of the chimpanzees of Kibale to ensure the *fusion* of their fusion-fission society. It is noteworthy that even in Gombe, where territorial killings reached their brutal peak, *twelve* years passed before the first confrontation between males was witnessed, and subsequent to that, rarely more than one serious attack was witnessed by any member of the research team each *year*.

By their nature territorial clashes must be rare. Otherwise, with the extremely low birth rate of wild chimpanzees, they would vanish through self-destruction. In fact, I suspected that lethal territoriality resulted not so much from males of opposing communities meeting along boundary areas and, in the excitement of the confrontation, losing control of their instincts for self-preservation by leaping into a lethal melee, but by cold opportunism. It seems far more likely that, over the months, or even years, males casually monitor the presence and numbers of their competitors beyond their territorial boundaries (which Nishida and Kawanaka [1972] found to be *social,* not geographical phenomena and Goodall [1986] found to be extremely unstable), and decide to be the aggressors in combat only when the opposing side appears unmistakably inferior in numbers. What makes chimpanzees so fascinating is that, even though territorial *wars* are extremely rare, the males here in Kibale Forest consistently behaved with such solidarity toward one another as to appear *prepared* for war at any time. So, in conclusion on the question of whether territoriality is the standard pattern of wild chimpanzees, *unprovisioned* chimpanzees, the data from Kibale suggest it is. But I must still admit I was hoping to witness a serious clash.

While writing this chapter and mulling over the role of inference in science, I happened to also be reading *Comet* by Carl Sagan and Ann Druyan. Not only is the entire book about comets, it describes in detail the tremendous role the Oort Cloud, the source of all comets, has played in the formation of our solar system, and of the universe as we see it today. You can imagine my reaction to the following admission (p. 201), "Many scientific papers are written each year about the Oort Cloud, its properties, its origin, its evolution. Yet there is not a single shred of direct observational evidence for its existence." Suddenly I felt much better. After all, I had *seen* the apes of Kibale behave *exactly* as predicted by the sociobiological hypothesis explaining cooperative territoriality by kin groups of male chimpanzees as a logical result of sexual selection pressuring a communal reproductive strategy based on inclusive fitness. Yet, in the back of my mind, I still wondered, against all odds and logic, whether the Kasakela Community of Gombe was unique in their lethal attacks. Were they after all an aberrancy?

Then, with almost uncanny timing, Toshisada Nishida sent me a report (1985) by him and his colleagues from a German journal concerning the Mahale Mountains Project. During the second decade of observation all six adult males of the smaller community, K-Group, containing twenty-two members, vanished one by one, apparently due to aggression by the males of the other two much larger neighboring communities, who were dominant because of their size. One after another the females of the annihilated community transferred with their surviving offspring (with the notable exception of most of the adolescent males, a few of whom actually wasted away and died) to the victorious M-Group, who also annexed much of the defeated K-Group's territory. Among other things, Nishida and his colleagues concluded, "This observation effectively rejects the males-only community model of chimpanzee social organization." One of those other things was fairly obvious. By becoming killer apes, the males of M-Group had just gained much more of the two resources which most limit the reproductive potential of males: adult females and good territory in which to raise offspring.

The only other species whose society closely resembles that of the male-retentive, cooperative, traditionally territorial chimpanzees (other than their sibling species, bonobos) is man. In both species males form close companionships with one another, sometimes hunt or defend territory cooperatively, mate with women who immigrate into their closed societies (G. P. Murdock classified most human societies this way), and display a keen sense of territorial belonging. Primitive hunting and gathering societies the world over exhibit, or once exhibited, patterns of exogamy, hunting, and sometimes ecology, plus territorial defense and warfare basically identical in form *and* function to that of chimpanzees. And, although chimpanzee warfare is still in the Stone Age (antagonists at Gombe sometimes rolled boulders down slopes at adversaries during confrontations), this should not mask the fact that, in Tanzania it was true warfare. Jane Goodall (1986:530) admitted that the violence was so unambiguously deliberate and brutal that, "If they had had firearms and had been taught to use them, I suspect they would have used them to kill."

I am not attempting to represent chimpanzees as humans, but similar adaptations of each species have led to amazingly similar social systems. Or have they? Chimpanzees and bonobos are our closest living relatives and probably have been for a good portion of human and protohuman history. The DNA-DNA hybridizations by Charles Sibley and Jon Ahlquist (1984) indicate that chimpanzees and humans diverged from a common ancestral lineage approximately within the last eight million years. And only a few million years prior to that gorillas diverged from our lineage. Because the societies of the three most closely related hominoids living today (chimps, bonobos, humans) retain their males in their natal social groups but lose their females to emigration (a system directly opposite to that among the Asian homi-

259

noids, the orangutans, plus almost every other nonhuman primate), it seems very likely that our present commonality is due to the African hominoid ancestor's somehow evolving this rare reproductive strategy of retaining kin lines of males. If so, and this seems quite likely, we share an unusual genetic legacy of cooperative male aggression against alien males.

And, although it would be comforting to conclude that being human places us above the kind of savagery evinced by the chimps of Kasakela Community, our track record reveals that we are far worse. Does war run in our genes like addictive behavior, diabetes, and baldness? Or does it run even deeper? Or could our legacy of war be merely cultural? My opinion is clearly that it is genetic, but, all sociobiological debating aside, I also feel we are intelligent enough to override this legacy. But are we responsible enough?

In view of the current scorched earth offensive that humanity is launching against the nonhuman primates of this planet—destroying their habitat at truly astonishing rates for commercial profit (Chapter 5) and capturing and killing tens of thousands of primates each year for medical experiments (next chapter)—we should be thankful that chimpanzees have not learned nuclear physics.

At dusk late in my vigil at the huge *dawei*, I slowly turned north from 8.5 to walk up C.5 for a change in perspective as I headed toward camp. Being preoccupied with some of the observations I had made that day, I allowed my attention to the path to lapse. When I looked up from a tangle of roots on my route, I froze. R.P. was sitting on the trail only a few yards ahead. He gazed at me from the corner of his eye, then scratched his arm briefly. I stopped and squatted down too, then scratched my left arm conspicuously and wondered what to do next. I felt my emotions suddenly jump gears into overdrive. Meeting him like this was something I had fantasized about for more than a year.

Here was my main liaison among the apes. Although his effect was usually subtle, R.P.'s nonchalance in my presence had worked so many times as a catalyst to calm his companions when they were nervous, that he had unknowingly become an important asset to my research. I felt I owed him some sort of concrete thanks. And, furthermore, even though many of the chimps were now tolerant, R.P. was the only one who seemed to take a perverse pleasure from my proximity. A pleasure that was reciprocal. He glanced at me briefly, then looked away. He continued to sit and betrayed no sign of unease. I was afraid of pushing my luck with him, but opportunity was knocking.

He scratched himself with a sound like sandpaper, then tentatively, probably unconsciously, patted next to his back wound. I thought about the salve in my pack and reminded myself that using it on his wound was only a fantasy, a one-in-a-million chance of thanking him for all the times he had helped me. But now the fantasy looked possible; only a few yards separated us. Would he allow me to approach him?

The question tormented me. But, like Bess, I was afraid to leap the gap, though I wanted to more than anything else. I would have given anything to communicate clearly and precisely to R.P. that I was his ally. After nearly half a minute my heart was hammering. I slipped off my pack, reached into it for the salve, then turned to glance at him again. He looked back, and I had the impression that he shrugged, not as a human does but as an ape who has made a decision to give up on something. Then, before I reached my feet, he turned and melted into the foliage on his way to find a good tree for a night nest. Instantly I cursed the caution that had made me miss this chance.

The end of this vigil gave me time to appreciate all my opportunities to observe the apes of Ngogo. I was acutely aware that my time was running out and my path would soon lead from the forest. Now, as I looked again across the swamp at the golden sunlight playing on the leaves of the forest giant, no chimpanzees were in it. The now empty *dawei* seemed symbolic of Africa's forests of the future.

I was indulging myself by dwelling on how much I was going to miss the apes of Ngogo. Owl's clumsy style, her greeting of me; R.P.'s nonchalance and trust; Butch and Lysa grappling high above the forest floor with fearsome grimaces on their doll-like faces; ancient Gray smacking her lips while immersing herself in the even older ritual of mutual grooming; the calm, inquisitive gaze from Stump's sunken eyes; Eskimo's muscle-bound displays and exaggerated aloofness over the petty affairs of lesser chimpanzees; Bess's hesitation at the end of the gap; Felony's phony grooming; the unearthly pant-hoots and throbbing buttress-drummings; mouths so stuffed with figs that each chewing spills out a little more into a waiting hand; and, last, the subtle turning of the head after gazing at me, the chimpanzee equivalent of a shrug, as if to say, "He's harmless." My reverie was self-indulgent because what the chimpanzees really needed were active allies among the alien invaders, mankind.

In fact, my continued efforts at convincing them to lower their guard, their natural fear that serves them in good stead against such a remorseless enemy as man, now gave me cause for regret, because it may have increased their vulnerability. Being their only liaison with *Homo sapiens* devolved on me a responsibility to help secure their future. Chimpanzees are vital in the vanishing forest that has been living since before the first man was born, but now the apes were under attack—and apparently were losing.

15

Cain and Abel

Millions of years ago, in an age when chimpanzees may not have been our closest phylogenetic relatives, the ancestral stocks of chimpanzees and humans evolved together in Africa. As the earliest hominids (Hominidae is the toxonomic family of humans and protohumans) tentatively took their first bipedal steps into the savanna woodland, possibly a new world for higher primates, primitive chimpanzees already may have been accomplished specialists of fruit in the vast tropical forest of those times.

As the forest shrank and the savanna expanded, those early hominids evolved rapidly into the splendidly versatile, but physically unimpressive beings we consider protohumans, and later true humans. First as scavengers, then as hunters of increasing skill, their niche took advantage of the abundance of meat roaming the ancient grasslands to augment their apelike opportunism in diet. Early humans probably fell prey to carnivores and probably competed with them to some extent, but they survived on a modest scale and managed to evolve into an ecological jack-of-all-trades hunter-gatherer. Eventually they cleverly exploited nature to become an agricultural technocrat that now threatens the stability and future of most ecosystems on earth. In the process that left their frugivorous cousins, the chimpanzees, behind evolutionarily in their narrow forest niche.

The ecology of chimpanzees is dominated by the exploitation of ephemeral patches of fruit in a competitive environment. The foods of the apes are rare. Their fusion-fission, male-retentive society is even rarer. In fact, were a biologist to theoretically design a creature with all the attributes of chimpanzees and then propose its existence to a primate ecologist, the latter would assure him that not only was his creation improbable, it was impossible. If chimpanzees did not exist as they do, we would consider it impossible that they could, especially against the competition they face in the forest. But their intelligence and versatile fusion-fission society has assured their success—until now. Without the forest, however, they cannot exist.

Before the combination of Western agriculture and medicine led to the current population explosion of *Homo sapiens* in Equatorial Africa, chimpanzees held their own in the forest. Human hunters shot them with poisoned arrows, speared them, snared them, and netted them, but the chimpanzees were intelligent, hunters limited, and the forest vast—so the apes survived. Now Africa is exploding with hundreds of millions of the modern version of that ecological jack-of-all-trades primate of the savanna and they require continually more food and more land on which to grow it. The mature tropical forest, once apparently more inexhaustible than the prairie-bison ecosystem of North America, is vanishing and with it our specialist cousins. In an evolutionary version of the story of Cain and Abel we are murdering our phylogenetic brother, the chimpanzee.

Droplets of dew spilled from the foliage as she brushed against it. She was stiff from having spent more than eleven hours in her nest, but the night had been comfortable and she was dry. It was January (1978), the height of the dry season. Her stomach rumbled with hunger but little fruit was available. She spotted a clinging vine, stopped, and plucked the plant's little photosynthetic factories one by one and ate them. But the leaves were hardly more than an appetizer; the hollow sensation in her middle was undiminished.

Not far ahead the forest ended. Beyond that abrupt termination began an apparently limitless open expanse, the domain of humans. Only ten years before, when she had ridden secure on her mother's warm hairy back, the old female had sometimes crossed those areas of strange plants to feed on the fruit of some isolated trees when they produced. She dimly recalled those days, but she would have no trouble retracing her mother's steps to those trees; she had walked the route several times herself while following her mother and younger brother, who had usurped her riding position.

But those isolated trees had not produced fruit for many seasons; they had been cut for timber and firewood. And her mother had not traversed the fields for many months because she was dead. The forest itself had changed as she grew; many of the large trees had been chopped down and others were now dead from arboricides. Black and white colobus monkeys abounded in the lush new growth shooting into the sunlit gaps caused by the deaths of so many forest patriarchs. The forest of her infancy was not the forest of this day, but she did not know why.

She was not fully mature and not pregnant. Often she traveled alone. Food was sometimes hard to find. But today she knew the location of an abundance of sweet food. For several days now she had passed beyond the boundary of the forest to visit a cultivated field where tall sugar cane soon would be harvested.

It was still so early that only the first glimmerings of dawn had diffused to pierce the shadows on the forest floor. She had to reach the field early and feed hastily to avoid the humans who would chase her away when they emerged from their huts with eyes bleary and red from wood smoke and crude banana gin. They frightened her, but her hunger and the sweetness of the cane drew her back.

The knuckles of her hands left shallow impressions in the earth as she moved silently along the game trail toward a section of the field where she could not be seen from the huts. Now she was so close that her memory of the sweet taste drove her to anticipate it anew and her mind wandered.

As she plodded along putting one hand in front of the other, the surface of the path abruptly collapsed and her hand sank below the surface. Simultaneously a thick sapling snapped upright with a frightening sound, and something constricted her wrist and pulled it to one side of the hole. Panicked by the noise and the strange thing that had gripped her, she tried to yank her hand out of the hole. The desperate power of her attempt cinched the thick braided wire into a tight loop around her wrist. Her heart pounded as she tried again and again to extricate her hand, but all her efforts served merely to tighten the rusty loop more painfully. The snare was well constructed.

Pure fear crawled within her as if a separate organism. She inserted her face into the trench and tried to bite at the wire but to no avail. Picking at it with her free hand proved useless too, but she continued to try. She screamed in fear and frustration, then whimpered. She did not understand how or why she had been trapped but she knew she was trapped, and all her instincts told her she was in danger.

A few Batoro heard her screams from their *shamba*. The owner of the field guessed that another *sokomutu* (literally "person of the market" but meaning chimpanzee) was caught, then he smiled. Soon he sent one of his children to inform the professional poachers whom he had asked earlier to set up snares to catch the crop raiders. Only a few weeks ago they had caught an adult male *sokomutu*. Experience had taught the poachers that it was best not to allow a trapped chimpanzee to remain in the snare too long; sometimes they tore the wire loose and escaped.

Later that morning men with spears arrived at the *shamba*. The owner greeted them cordially; he was pleased that they were having success. Poachers had learned that net drives were not the best way to kill chimpanzees. The apes were too large and formidable; they sometimes attacked and killed their dogs and escaped the nets. Snares were better. The spearmen entered the forest.

Long before she saw them she heard them coming. Her anxiety soared as the sounds of their approach waxed. She redoubled her efforts to yank her hand from the pit. The wire bit through her skin and into her flesh. Blood dribbled down her hand to spatter the litter of leaves in the hole. She looked up and saw the spearmen before her. Her bowels loosened and fear flooded her.

The spearmen were careful to keep their dogs away from her. They respected her strength and there was no need to risk the dogs. Their concern stemmed not from solicitude for the canines, but out of regard for useful property, which the dogs were during net drives. Because they had no need for haste, the first man unhurriedly studied her to determine how he might best spear her in a vital place with one thrust. He was careful not to get too close to her while she was still alive. He had no wish to be bitten.

Surrounded by enemies and unable to run, she was lost. She pulled her arm again but it would not come free. Sugar cane no longer was important; the only thing now was to escape and live. The spear blurred like a streak of light. It entered her side, cut between her ribs, sliced through her lungs, and felt like fire.

Seeing her convulsive reaction, the other two hunters threw their spears into her too. Then the spearmen patiently waited for her to bleed to death. Occasionally she jumped or convulsed, still trying to escape and dislodge the spears. These hopeless attempts brought smiles to the men's faces. But as her attempts grew more feeble and their novelty waned, they afforded less amusement. In a few minutes she was dead.

One of the men kicked her to ascertain her condition. Then they butchered her. They considered her flesh unfit for human consumption but it was good for their dogs. One man decapitated her with his *panga* and, with the iron butt of his spear, pried loose the occipital bones from the base of her skull to enlarge her foramen magnum. He scooped out her still warm brain and called the dogs to eat it so their coats would become silky.

He saved the skull. After butchering her for dog food they took the skull and some of her other parts to sell to a shaman in Kanyawara who specialized in marketing such items. The skull is believed to possess curative powers. Its most frequent application is for healing broken bones. The skull is burned to charcoal, then pulverized to a powder and mixed with other ingredients to form an expensive paste. Incisions are cut in the skin of the patient and this paste rubbed into them, thus effecting a putative speedy knitting of the fracture. The skull of the male killed in December had already been used for this purpose and had earned the shaman the equivalent of a month's wages.

A Ugandan informant gave me the details of this incident. The female chimpanzee was snared on the edge of the forest between Nabinamba and Lwamugonera, near the sugar cane plantations about three miles north of Kanyawara. The poachers were under contract to the cane growers. I examined her skull. She was about as old as Owl or Ita; her wisdom teeth and permanent upper canines had not erupted fully. While the fine detail of my account might omit something from the sequence of events that occurred, or might vary from it some, the basics of the situation remain the same: chimpanzees in Kibale Forest were being killed commercially.

Because Batoro do not eat chimpanzees, the hunting pressure on their population in Kibale is much less than what it might be, and what it used to be when Bakonjo hunted there. But killing them at all is illegal, unnecessary, and, in my opinion, immoral.

During my final months in Kibale Forest I thought hard about the apes needlessly snared, amputated, or speared to death in the forest without a single obituary to alarm a concerned person. I eventually saw ten chimpanzees, or their remains, who had been mutilated and crippled or killed like this. Meanwhile the forest itself was being quietly and illegally whittled away around its periphery by a gnawing horde of insatiable humanity. Within the Forestry Department's legal procedures, inherited from the British foresters of the Uganda Protectorate who were working hard to make Uganda pay a fair fee for Great Britain having taken up the White Man's Burden, parts of the forest were being felled on a seventy-year cycle or arboricided to remove undesired species. Both these processes degraded or totally destroyed the forest as a viable habitat for many of its inhabitants (see also Chapter 5).

Although relatively rarely victims of poachers, the apes of Ngogo were not immediately threatened by habitat destruction because at least half their range was protected, by *custom* (not law) within Kibale Nature Reserve at the core of the central block of the forest. The bulk of Kibale Forest, however, is simply a nationally owned resource whose future exploitation was planned by the British and still stands. This felling plan seriously threatens the Kanyawara chimpanzees and most of the other apes which keep their population genetically viable. They needed the *entire* central block. Now the responsibility of being their sole liaison with the lords of the planet spurred me to find a way to help conserve them and their forest, to make their situation known. I could not return to the U.S.A., write my scientific report, graduate, and turn my back on their future. They had become a part of me.

I had to admit objectively that the chimpanzees of Kibale were merely one more victim of the threat of encroaching humanity.

And they were not as badly off as some of their ape cousins. All of hunanity's close relatives who have managed to trudge down an evolutionary path to the twentieth century are facing the same specter. All are obligate denizens of tropical forests. At least the situation in Kibale Forest is promising because exploitation is slow and because groundwork by Tom Struhsaker, Tom Butynski, Joe Skorupa, Karl Van Orsdol, myself, and others has provided an good data base for planning a conservation strategy.

Smaller species generally survive better under man's onslaught then their larger cousins. In terms of numbers, the half dozen species of lesser apes are doing best. Gibbons inhabiting the shrinking forests of the Indo-Malay region (Southeast Asia and Indonesia) may number nearly 800,000 according to Jaclyn Wolfheim's *Primates of the World*, although this is a drastic decline from their numbers of only twenty years ago. Deforestation plus hunting for meat and the commercial pet trade are the primary causes of their decline.

In the mid-1960s the future of the most primitive and arboreal of the great apes was considered in imminent peril, mostly due to hunting and partly to logging. Multiple field studies in Indonesia since then have revealed a slightly rosier picture. Herman Rijksen (1982) estimated 5,000–15,000 orangutans remained on Sumatra. Birute Galdikas, John MacKinnon, Peter Rodman, and other primatologists found healthy populations across Borneo, where orangutan habitat is about six times more extensive than on Sumatra and may contain a total of twice as many apes.

But the threat of habitat destruction is now grave. Norman Myers (1984) reported that in 1980 Indonesia earned more than two billion U.S. dollars from timber, more than from all its agriculture and only second to oil revenues—and the rate of logging continues to accelerate. Indonesia not only has cut more than half her forests, land once forested but now useless exceeds nearly half a million square kilometers, 20 percent of the entire nation! Orangutans are protected in some reserves and a couple of national parks. During an exploratory expedition down Sumatra's Alas River I was able to visit their largest and most critical stronghold, Gunung Leuser National Park (3,000 square miles of forest). Unfortunately, most other refuges are becoming islands surrounded by rising seas of humanity and ecological devastation, and these refuges had no guarantee against being logged for commercial sale to escalate Indonesia's material standards to those of the Western world.

Some of the African great apes are better off. In West Africa their primary habitats generally are in nations less populated and less hurried about converting their primary forests to foreign exchange income, but, unfortunately, they are more inclined to *eat* great apes. Good data are scarce and, although agreement exists that all populations are declining, just how many do survive is undertermined. Jaclyn Wolfheim estimated 10,000 western lowland gorillas survive in Africa, but a recent nest-count census (patterned after mine in Kibale) done by Carolyn Tutin and Michel Fernandez generated an estimate of about 35,000 lowland gorillas in Gabon alone, which is their primary stronghold, mostly in undisturbed secondary forest. In several regions of Gabon, however, gorillas have been exterminated by hunting. Gorillas were reduced in areas of heavy and light hunting by 72 and 17 percent, respectively.

Eastern lowland gorillas, separated in extreme eastern Zaire 600 miles from their western population, are far less numerous: perhaps 2,500 altogether. Like orangutans they are threatened most by habitat destruction, but also secondarily by hunting for meat, body parts for magic talismans, in retaliation for crop raiding, and to capture infants for sale. Even so, their population is much less precarious than that of mountain gorillas.

Dian Fossey (1984) reported surveys in 1981 by a coordinated team of researchers who counted 242 mountain gorillas in the Virunga Volcanoes of Rwanda, Uganda, and Zaire—a 50 percent decline from George Schaller's estimate of twenty years earlier (1963). Although hunting and snare trap deaths contributed to part of this decline, probably the main factor responsible was habitat destruction. When I visited there during this survey, Conrad and Rosalind Aveling of the Mountain Gorilla Project were working to increase touristic income and to educate local people in the natural history of gorillas and the conservation value of the park. They explained ot me how in 1968, Rwanda, with the greatest population density of humans in Africa (512 people/mi² in 1980), presented approximately 38 of 90 square kilometers of Volcanoes National Park to Rwandese peasants in a 50:50 deal. The low-lying, productive acreage of the mountain gorillas' last retreat became available for clear-cutting and cultivation on condition that every acre planted with other crops had to be matched with equal acreage planted with pyrethrum, which produces a natural insecticide. In addition to hurting mountain gorillas, this deforestation seriously impacted the watershed function of the park. Some perennial creeks stopped flowing during dry seasons.

The only other population of mountain gorillas, a small but unstudied one, lives in the Impenetrable (Bwindi) Forest of southwest Uganda. Tom Butynski and Jan Kalina have recently started a project there to determine its ecology and status. At present Butynski is hoping they may number nearly as many as in the Virungas, but many more censuses are needed to be confident. (I have been trying to initiate a similar project on the chimpanzees there, but so far have not gained funding support.) If the Bwindi gorillas are actually mountain gorillas, they are the only mountain gorillas to share their forest with chimpanzees.

Bonobos (pygmy chimpanzees) are restricted to forests of central Zaire confined by the Congo, Zaire, Lualaba, and the Kasai and Sankuru rivers. Using the few available reports, Jaclyn Wolfheim estimated 100,000–200,000 bonobos may remain in a declining population, but more recent researchers consider this a serious overestimate. More and better censuses are vitally needed. Though substantial, bonobo habitat destruction is less intense than for other apes, but may increase soon due to a joint Canadian-Zairean project for major timber extraction in their range. In most places where they occur bonobos are illegally hunted for food, magic talismans, reprisals for crop raiding, and to capture the infants for pets or for sale. Despite an entire symposium volume on bonobos edited recently by Randall Susman, much remains to be discovered about them—it is even questioned whether they exist within any national park. Their plight is ripe for preventative conservation.

Chimpanzees still occupy the greatest range of any great ape. Though now shattered into many small island fragments which probably contain too few individuals to support genetically viable populations, their current range extends from southeastern Senegal, where William McGrew, Pamela Baldwin, and Carolyn Tutin found a sparse population of only 0.23 apes mi^2 (only 5 percent of that in Kibale) in the hot, dry, open habitat around Mt. Assirik, and from southwest Mali where Jim Moore briefly surveyed a

271

dwindling population which may be even more sparse, east for thousands of miles to the riparian woodlands of western Tanzania. Paradoxically, so many studies of chimpanzees have been done, though often with only a few months' investment, that disagreement over what might constitute their preferred habitat is rampant.

Adriaan Kortland's (1983) painstaking ecological sleuthing had revealed that chimpanzees still occupy *several* dry, quite marginal habitats, which is convincing testimony that they may be the most adaptable of the great apes. Some writers even claim that human alteration of natural habitats may make them more suitable for chimpanzees, which, unless this is intended to mean providing agricultural fields for them to raid, seems ignorant, or intentionally misleading. It is obvious now that chimpanzees have a much wider habitat tolerance than other great apes, being opportunistic but still depending on the presence of mature, fruit-bearing trees. It is revealing that the highest densities for chimpanzee populations have been recorded in Uganda's Budongo and Kibale forests.

But a reliable view of their population sizes is impossible due to a serious lack of adequate censuses. Jaclyn Wolfheim summed it up thus: "Chimpanzees are thought to be abundant in parts of five countries, decreasing in six, increasing in two, and extinct in all or parts of four." In eight of these countries no recent data are available. Wolfheim quoted Adriaan Kortlandt as saying in 1974 that chimps are "gravely declining almost everywhere in the entire range from Sierra Leone to the Kivu Mountains." With possibly a few exceptions, this is almost certainly true. Countries believed to have populations exceeding 5,000 chimpanzees include Guinea, Cameroon, Zaire, and Gabon. But the only reliable census available is from Carolyn Tutin and Michel Fernandez's nest counts in Gabon, which when extrapolated with certain positive assumptions, indicate as many as 64,000 individuals survive there, a very encouraging estimate. Geza Teleki (in press) has provided the most up-to-date summary of chimpanzee survival on a nation-by-nation basis in Africa; the overall picture is alarming. Teleki's best-guess estimate is that perhaps between 40,000 and 100,000 chimpanzees remain in Africa.

One unequivocal trend is the serious reduction of chimpanzee populations after arboricide refinement or selective felling, and their total extinction after normal logging operations. Because of accelerated logging by foreign contractors in Africa, habitat destruction is the greatest threat facing chimpanzees today. But, again, despite its illegality, nearly everywhere that chimpanzees survive outside national parks (seventeen of them) they are hunted by all methods, including firearms, for meat, skulls, and other body parts for fetishes and magic, in retaliation for crop raiding, and to capture infants to supply the local and international demand for pets, zoo animals, and especially biomedical research subjects.

Jaclyn Wolfheim reported that, according to the Interagency Primate Steering Committee of the National Institutes of Health, the U.S.A. requires 350 chimpanzees each year (among the 33,912 primates of all species "used" each year in the U.S.A.) for experiments on diseases such as AIDS and hepatitis B. A current moratorium on importing chimpanzees into the U.S.A., however, has limited new use to the seventy-five or so born yearly in captive breeding colonies totaling roughly 1,100 chimpanzees.

These captives are far less successful reproductively than wild chimpanzees because of the unnatural and often bizarre psychological environments of colonies. Captive-born chimpanzees, even among the 400 in U.S. zoos, almost never reproduce successfully. Current methods of capturing wild infants are brutal in the extreme. Geza Teleki (in press) reported that the unsupervised (i.e., usual) capture of infants normally entails killing the mother and, as among gorillas, possibly other group members. What's worse, Teleki computed that for every infant delivered alive, five other captured infants die through wounds, mistreatment, and/or psychological trauma before their "safe" delivery within a developed country. And because the mother of each infant has been killed, ten chimps actually die for each survivor delivered. Hence, *thousands* of chimpanzees have been murdered annually for the infant trade alone. Carolyn Tutin and Michel Fernandez found that light and heavy hunting of chimp populations not already extinct in Gabon reduced the chimpanzees by 24 and 57 percent, respectively. It is not only a jungle out there, it is war.

If "war" sounds exaggerated, consider Ursula Rahm's report (pp. 202–3) of the joint, *supervised* efforts in the mid-1960s by the Institut pour la Recherche Scientifique en Afrique Centrale

(IRSAC) and the Delta Regional Primate Research Center (of the U.S.A.) to capture wild chimpanzees on forty occasions in the lowland rain forest of eastern Zaire immediately west of the Mountains of the Moon. First, scouts followed the chimps until they built nests for the night, then eighty natives surrounded the apes at midnight with a quarter mile of nylon netting.

> Then, at dawn, when they climb down to the ground, beaters inside the net are pushing them towards the nets, shouting and drumming. If they feel and fear the human presence they won't descend to the ground and try to escape over the trees. If they panic and rush into the net, they are immediately enveloped in its baggy base . . . At this moment they are immobilized by a series of sticks planted around and crossed over them. . . . Then they get a Sernylan injection (Phencyclidine hydrochloride) and usually are sleeping within 5–7 minutes. Adult females are frequently the ones caught first and panicking most. Animals aware of the net investigate it from a distance of about 3 meters and disappear quietly into the undergrowth to hide sometimes for up to four hours. They stay motionless even if the beaters pass within two meters of them, then suddenly they rush toward a small tree or banana stem and swing themselves over the net. Only the stench of their diarrhea caused by fear indicates that they must be very nervous. If ever the animals see a possibility to escape over the trees, they do it without hesitating. Very often they run upright like gibbons over the branches and jump distances they would not try under normal conditions. The only fatal accident in a chimpanzee we ever saw during the captures was an old male underestimating the distance he was jumping from one tree to another and falling about 15 meters.

Fifty-six chimpanzees were captured this way, most of whom were destined for lab experiments.

To maintain perspective one must bear in mind that the description above is of *supervised*, relatively humane captures—a procedure that no longer occurs. Chimpanzees now captured for the biomedical research market are caught by local hunters who shoot the mother (many also eat her).

Although a dismal picture, it certainly is not hopeless. On paper chimpanzees are legally protected almost everywhere. Unfortunately, most African nations in which they live have many other priorities in law enforcement they consider higher than conservation. Many recommendations have been made by scientists to preserve essential habitats (for example, Vernon Reynolds's (1965) recommendation to not cut *Ficus mucuso* trees or use arboricides in Budongo Forest), and occasionally the forestry departments listen. But they earn little money in the short term by practicing conservation, and Africa's pragmatic perspective these days is short term, if not immediate. So incentives to conserve appear to many of them to be based on altruism rather than the enlightened self-interest they really are. Tanzania has been exemplary, having created several national parks including the Gombe Stream Reserve and the Mahale Mountains, despite serious national economic problems. Doing the same with Kibale Forest Reserve would be an even more farsighted move to preserve an irreplaceable national treasure of Uganda because the forest is so much larger and richer then the Tanzanian parks containing chimpanzees. It would also help to ensure the survival of a genetically viable population east of the Ruwenzori Mountains.

Meanwhile, to make the most efficient use of the captive chimpanzees now in the U.S.A., the National Institutes of Health are attempting to determine how to improve the psychological health of captive-born chimpanzees in order to turn them into breeders. Examples from this book, and from the books of Jane Goodall and the many reports of other scientists studying chimps in the wild, describing the freedom and rich social lives wild chimpanzees enjoy, provide insights as to how difficult a challenge it is to duplicate psychological health in captivity. Difficult but not impossible. The chimpanzee colony living in the spacious natural enclosure in the Burger's Zoo in Arnhem, the Netherlands, reported by Frans de Waal, proves the goal is realistic. Despite the fact that it can be done, the changeover from the medieval dungeon concept, updated by electric lighting, common in most primate research facilities, to the open, spacious acres of a community enclosure is expensive. More expensive than simply paying U.S. $5,000–10,000 for an infant chimpanzee torn from its dead mother in the wild, then perhaps slipped through a bureaucratic backdoor to avoid interception by customs officials.

In their Policy Statement on Use of Primates for Biomedical Purposes, the World Health Organization (WHO) and Ecosystem

Conservation Group recommend "endangered, vulnerable and rare species be considered for use in biomedical research projects only if they are obtained from existing self-sustaining captive breeding colonies" (i.e., in captive breeding, all animals are required to be at least F2 generation). They further recommend, "wild caught primates be used primarily for the establishment of self-sustaining captive breeding colonies, the eventual goal of which should be to captive breed most or all (depending on species) of the primates used in research" (see McGreal, 1987).

Despite these recommendations, the biomedical research community in the U.S.A. and Europe continues today to lobby aggressively to end the moratorium on importing more chimpanzees from the wild, an importation proscribed by the Convention on International Trade in Endangered Species of Wild Fauna and Flora (CITES), despite the U.S.A. and most European countries having signed the Convention. Such lobbying can take many forms, some of which would be termed something else by purists. One Austrian-based company with an avowed continuous need for several chimpanzees annually for hepatitis and AIDS research was engaged recently in some "embarrassing" lobbying to circumvent Sierra Leone's 1979 ban on exporting chimpanzees and Austria's signing of CITES. To avoid exacerbation of their embarrassment, the company sued Shirley McGreal, chairman of the International Primate Protection League (IPPL), the journal *New Scientist*, J. Moor-Jankowski, editor of the *Journal of Medical Primatology*, Austrian journalists, the president of the Vienna Animal Protection Society, a professor at the University of Veterinary Medicine, and nearly fifty others for allegations of illegal action difficult to prove. At the same time they started negotiations to construct a research center *in* Sierra Leone, which appears likely to be built, so as to experiment on chimpanzees without bucking the export and import restrictions. McGreal and many other scientists fear the disastrous repercussions of "veterans" of hepatitis or AIDS rehabilitated back into the wild—as the company actually planned at one stage—where they would spread the disease throughout the wild populations of Sierra Leone. (More on these sleazy corporate machinations can be found in Ian Redmond's special report, *Law of the Jungle*.)

Although biomedical lobbyists recommend that captures be made only by trained personnel and preferably in areas where the apes are doomed by their host governments having awarded their forests to foreign logging concessions, this is merely a humane smoke screen to procure chimpanzees quickly and at the cheapest possible price. There exists no way to verify the origin of a captive infant chimpanzee. I find the prospect of further captures, legal, quasilegal, or illegal, abhorrent to put it mildly.

This raises an important issue. Most people agree that no one has the ethical right to destroy any of the nonpathogenic species on earth, both because it denies future generations of people their natural posterity and because it divests the ecosystem of living diversity. Legal shades of gray are encountered, however, in circumstances wherein an endangered or threatened species is "merely" being willfully hastened into the black hole of extinction by taking almost, but not quite all of the remaining members of the species. Chimpanzees are classified as "threatened" by the U.S.A. Endangered Species Act, as "vulnerable" by the International Union for the Conservation of Nature and Natural Resources, are included in Appendix I of CITES, and are considered substantially threatened in the analysis of Jaclyn Wolfheim.

Most rational people also agree that no one has the right to murder an innocent *Homo sapiens*. For some reason, innocent members of other species seem to be an another issue entirely. Why? Because humans are special: they alone possess self-awareness and are rational thinking beings. For as long as *Homo sapiens* has remained in ignorance of the mental universes of other species, this belief had condoned lack of compunction in dealing with nonhumans (and even with other races of humans believed to be subhuman). But if another species were recognized to possess self-awareness and the ability to think rationally, would its members then also have the right to be protected from murder?

Sufficient reason exists now to conclude that we are not unique in self-awareness and rationality. One of the first hints of this came from Wolfgang Kohler's pioneer experiments during the first quarter of this century wherein he suspended a banana high beyond reach in a chimp's cage, or placed one beyond reach outside it. The ape in both cases had resources (boxes, two of which were needed to stand high enough to reach the banana, and hollow sticks singly too short to hook it) to solve the problem, but had no previous experience with this type of problem. In both cases, after a delay, chimpanzees had sudden "aha" bursts of activity and piled boxes on top of one another to climb to the banana. Kohler's brightest subject, Sultan, fitted sticks together to build one long enough to hook the banana outside. So?

Because of the novelty of the problems, the delays in coming to grips with them, and the sudden solutions, Kohler concluded the chimpanzees had demonstrated rational insight, defined by *Webster's Third New International Dictionary* as "immediate and clear learning that takes place without recourse to trial-and-error behavior." While most people might claim the solutions were far short of genius level, it is difficult to describe a mental process by which these solutions were attained without invoking insight.

Since Kohler's work, chimpanzees have been the subjects of hundreds of brain operations, psychological tests, problems, and training programs (well over one hundred chimpanzees, for instance, were trained for the U.S.A. space program). J. B. Wolfe and J. T. Cowles taught chimps to perform tests for tokens good for exchange for food and services (including being returned to the home cage). The apes learned their different values and saved up to thirty of them in anticipation of expensive purchases. C. B. Ferster taught chimps to count using a binomial system. Emil Menzel found that a chimpanzee who knew the locations of hidden foods but was restrained from getting them himself could communicate their hidden locations to other chimps.

Chimpanzees are also innovative. E. Hall reported a female, who, tested, by D. O. Hebb, repeatedly solved her test to win banana slices which she lined up in a row. When Hebb ran out of slices, she continued to solve the tests but paid Hebb a slice *back* every time she won. Hebb ended up with thirty slices. Several other observers have reported innovativeness in tool construction and use, both in the wild and in captivity. But the most dazzling displays by captive chimpanzees are with language.

Because the voice box morphology of chimpanzees precludes human vocal speech, Alan and Beatrice Gardner trained a one-year-old, wild-born female, Washoe, American Sign Language (ASL). Though erratic, her training was continued by Roger Fouts, Richard Budd, and several colleagues, and eventually Washoe used 170 ASL signs regularly. After being moved to a chimp colony she signed to the other chimps and once even to a snake, urging it to "come hurry up dear" as the other chimps fled from it. Eventually *she taught* ASL signs to a young orphaned chimp, Loulis, whom she adopted.

Ann and David Premack taught several other chimpanzees, including their star, Sarah, to communicate both abstractly and contingently by using more than one hundred plastic symbols. Duane Rumbaugh and Timothy Gill trained another young female chimpanzee, Lana, to communicate similarly via a computer keyboard. After years of training and tests, Gill concluded, "Lana clearly demonstrated that she is operating in a domain once held exclusive to man, that of language." Several other apes, including a star gorilla named Koko, have learned to use ASL extensively and appropriately since these initial experiments. Only one devil's advocate, H. S. Terrace, concluded on the basis of the poor performance of a single infant male, Nim, confronted by a bewildering array of dozens of instructors (with no reliable surrogate parent to provide natural motivation), that chimpanzees were not really using language but merely "aping" their instructors.

Why are chimpanzees so smart? Or, more to the point, *how* does being intelligent benefit them? High intelligence is indispensable for at least two basic reasons involving reproductive success: first, as I mentioned in Chapters 4 and 9, their feeding ecology depends on outcompeting many hundreds of monkeys for rare fruit when it ripens scattered throughout the several square miles of the apes' home range. The only way to ensure winning against this competition is to remember the locations of hundreds of fruit trees (exactly as I had to in order to find the apes), somehow mentally categorize the trees as to when they will produce, then consistently arrive at them first for the harvest.

The second, completely different demand for high intelligence hinges on the chimps' complex and protean social dynamics. Examples from the apes of Ngogo, plus many more from Mahale Mountains and Gombe (especially in Jane Goodall's recent and excellent book, *The Chimpanzees of Gombe*) reveal some of the almost bewildering social intricacies which wild apes create and with which they must contend. The imbroglio of interpersonal politics for which chimpanzees are truly capable is illustrated graphically by Frans de Waal for the captive colony in Arnhem, in which the apes are completely freed from any physical or mental challenge connected with finding food and, instead, indulge themselves (particularly the males) in constant machinations to form dominance alliances. These alliances seem aimed solely at monopolizing opportunities to mate. In the wild, similar alliances between kin, plus coalitions based on reciprocal altruism between nonkin or distantly related apes observed in Mahale and Gombe, are as Machiavellian as anything in a James Clavell novel. My impression is that high intelligence is vital for chimps simply to hold their own in the dynamic interplay of individual machinations within their community. It must be even more important in competition *between* communities.

Overall, the performance of chimpanzees in increasingly more sophisticated tests has revealed not only insight, arithmetic conceptualization, symbolic communication, abstract thought, and technological and linguistic inventiveness, chimpanzees in captivity and the wild have also revealed a sense of self-awareness and identity, awareness of the reality of death, lifetime bonds based on friendship, voluntary sharing of rare foods, knowledge of ethical right and wrong, the ability to lie to attain goals, plus an anticipation of the future. None of these mental capacities was seriously suspected to reside in the minds of apes, despite their brains being approximately the same size as those of the early australopithecines from whence we evolved, and despite chimps having a Broca's area, responsible for the structure and function of human speech, similar to modern humans. Would killing a living *Australopithecus* be murder? How about a *Homo habilis*? And now that we know so much about their mental life, how about a chimpanzee?

In 1947, Robert A. Heinlein examined the morality of this question in a piece of speculative fiction that was uncannily prescient. In the story a chimpanzee named Jerry, one of a group that had been modified genetically to allow vocal speech, was due for euthanasia because he was too old to be useful to the company who owned him. A stockholder became aware of Jerry's plight and tried to save him and his companions from this inhumane fate, and in so doing embroiled Jerry in a court suit to determine his status. The expert witness presiding over the court, an impartial Martian, ruled that because he could use human speech and *communicate* appropriately in the process, *Jerry was a man*. Killing him would be murder.

Exactly where *do* chimpanzees stand evolutionarily and genetically in relation to humans? The physical resemblance between us is so striking that it has led to endless comparisons ranging from the whimsical in Hollywood, to the pedantic in academia, to the putatively indispensable for saving human lives in the medical community. Several investigations of biochemical similarity by the teams of Mary-Claire King with A. C. Wilson, and V. M. Sarich with J. E. Cronin, by Dorothy A. Miller (and see the excellent summary of recent work provided by Nancy Tanner) agree generally that chimpanzees are so similar that they must have diverged so recently that traditional models of human evolution based on fossils (no chimpanzee fossils are known) seem to greatly overestimate the age of our common ancestor. Indeed King and Wilson came to the intriguing conclusion that "the genetic distance between humans and the chimpanzee is probably too small to account for their substantial organismal differences."

In the past decade Charles Sibley and Jon Ahlquist perfected a technique to compare DNA molecules (deoxyribonucleic acid, the basic genetic molecule containing the hereditary material in the nucleus of each cell in our body) of two species by physically splitting them, then matching them to determine how well they fit together. Their methods have revolutionized the classification of birds, their main target. But recently Sibley and Ahlquist also applied their state-of-the-art analysis to primates. They found that the African apes and man are much more closely related to one another than to orangutans or gibbons (which is in concert with the evolution of their social systems I proposed in the preceding chapter). Monkeys were so distant as to be out of the picture. But, contrary to one cherished model of evolution, the nearest relative of chimpanzees and bonobos (which differ genetically by 0.4 to 1.1 percent) are not gorillas (1.9 to 2.5 percent difference with chimpanzees) but humans (1.5 to 2.0 percent with chimpanzees). In other words, based on the genes themselves, we are the nearest living relatives of chimpanzees and bonobos! How is this significant?

Jared Diamond pointed out that cladistic taxonomy (based on genetic relations) may spur an upheaval in how we look at ourselves. His example begins with the common gibbon and the siamang, which taxonomists agree should be classified in the same genus, *Hylobates*. But Sibley and Ahlquist's work indicates that gibbons and siamangs diverged from one another longer ago and are genetically more different (1.9 to 2.5 percent) than chimpanzees and man. In short, to be consistent with state-of-the-art genetics, cladistic taxonomists no longer should refer to chimpanzees as *Pan troglodytes*, bonobos as *Pan paniscus*, gorillas as *Gorilla gorilla*, nor humans as *Homo sapiens*. All four species should be classified in the same genus, *Homo*. Chimpanzees should be *Homo troglodytes*. Do *Homo troglodytes* deserve ethical consideration? Should *Homo troglodytes* possess legal rights, such as protection from murder?

I am not trying to claim that chimpanzees are backward humans trapped in ape suits. But, based on what we know of their advanced mentality and close genetic relationship to us, neither are they on a level with the family dog. My conviction is that purposefully killing them is murder. Further, making their existence untenable by destroying their critical habitat is equally criminal. Aldo Leopold's Land Ethic (Chapter 5) may hold some hope for preserving enough tropical forest to ensure the survival of chimpanzees assuming we act before it is too late, but a change in attitude about the *status* of chimpanzees—and all the great apes—is long overdue.

On my last night at Ngogo I felt a painful rift. Learning about the chimpanzees had demanded such absolute dedication that I had been focused entirely on their lives and on the complexity of the forest. Where are they? Who are they? What are they doing? With whom? And why? The incessant searching, straining my senses to hear or see them, forcing my mind to anticipate their moves and to think as a chimpanzee does, had changed me. Now this break was traumatic. I understood that much of the emotion I was feeling was a compulsion to somehow alter the scenario I envisioned of their ultimate fall before the axes and greed of human civilization. This book seemed a step in the right direction.

Epilogue

Within the five years since I left Kibale Forest, two other projects on the chimpanzees there have been started. The first, now completed, was a postdoctoral project by G. Isabirye Basuta, a student (now a professor) from Makerere University working under Tom Struhsaker, on the Kanyawara chimps. I have not seen his report, but he explained to me that he studied their ecology. He also told me that, in hopes of expanding his scientific horizons, he tried to observe the chimpanzees at Ngogo but found, as had other more casual scientists before him, that they were far too wild. And, like the others, he gave up the idea as too difficult. In the past couple of years Richard Wrangham has started another, more detailed study of the use of understory plants as foods by the chimpanzees of the Kanyawara Community, a project that may shed further light on the complex ecology of our nearest cousins. But no one has followed up on the apes of Ngogo.

In November of 1986 the Chicago Academy of Sciences sponsored a first-of-its-kind symposium titled "Understanding Chimpanzees." This four-day event was spurred by Jane Goodall and gathered most of the scientists who have studied wild and captive chimpanzees during the last twenty-five years. While stimulating to the point of overkill, one basic consensus emerged from this symposium, one only tentatively connected to theoretical science: we concluded that chimpanzees were in extreme danger of being extirpated in most of their host nations, and that, if serious conservation work was not initiated on a large scale, a few decades hence any effort will have been too late. Most chimpanzees in the wild will have become extinct, and their remaining populations will have dwindled to a size too small to contain the genetic variability necessary for their survival.

Sobered by this specter, members of the symposium formed a committee, the Committee for the Conservation and Care of Chimpanzees (CCCC), which started its life under the aegis of the newly organized Jane Goodall Institute for Wildlife Research, Education and Conservation (P.O. Box 26846, Tucson, AZ 85726). Because of his recent success in conserving chimpanzees and directing Sierra Leone's first national park, Geza Teleki was named organizer and chairman of the CCCC. His first objectives are to seek endangered species status for chimpanzees under the U.S.A. Endangered Species Act and to raise support to begin censusing forests to determine, after all, where the chimpanzees' last strongholds are. Frans de Waal volunteered to begin work with the U.S. National Institutes of Health to upgrade standards of care for captive chimpanzees in the U.S.A. And Toshisada Nishida volunteered to begin cataloging those habitats remaining in Africa where the CCCC should send scientists to census chimpanzees. At present the CCCC's strategy for conserving chimpanzees is only a plan, not even a polished plan, and financial support must be gained before it may be implemented. But, for the first time, we are working together.

References and Background Reading

Albrecht, H. 1976. Chimpanzees in Uganda. *Oryx* 13(4):357–61.

Albrecht, H. and S. C. Dunnet. 1971. *Chimpanzees in Western Africa*. Munich: R. Piper.

Altmann, J. 1980. *Baboon mothers and infants*. Cambridge, Mass.: Harvard University Press.

Azuma, S. and A. Toyoshima. 1962. Progress report of the survey of chimpanzees in their natural habitat, Kabogo Point Area, Tanganyika. *Primates* 3:62–68.

Badrian, A. and N. Badrian. 1984. Social Organization of *Pan Paniscus* in the Lomako Forest, Zaire. In Susman 1984, pp. 325–46.

Baldwin, L. A. and G. Teleki. 1973. Field research on chimpanzees and gorillas: an historical, geographical, and bibliographical listing. *Primates* 14:315–30.

Baldwin, P. J., J. Sabater Pi, W. C. McGrew, and C. E. G. Tutin. 1981. Comparison of nests made by different populations of chimpanzees (*Pan troglodytes*). *Primates* 22:474–86.

Baldwin, P. J., W. C. McGrew and C. E. G. Tutin. 1982. Wide-ranging chimpanzees at Mt. Assirik, Senegal. *Int. J. Primatology* 3(4):367–85.

Bauer, H. R. 1975. Behavioral changes about the time of reunion in parties of chimpanzees in the Gombe National Park. *Contemporary Primatology 5th Int. Congr. Primat.* pp. 295–303.

References and Background Reading

———. 1977. Ethological aspects of Gombe chimpanzee aggregations with implications for hominization. Ph.D. dissertation, Stanford University.

———. 1979. Agonistic and grooming behavior in the reunion context of Gombe Stream chimpanzees. In Hamburg and McCown 1979, pp. 394–403.

———. 1980. Chimpanzee society and social dominance in perspective. In D. R. Omark, F. F. Strayer, and D. Freedman, eds., *Dominance relations: ethological perspectives on human conflict*, pp. 97–119. New York: Garland.

Beauchamp, G. K., K. Yamazaki and E. A. Boyse. 1985. The chemosensory recognition of genetic individuality. *Scientific American* 253(1):86–92.

Bertram, B. C. R. 1973. Lion population regulation, *E. Afr. Wildl. J.* 11:215–25.

———. 1976. Kin selection in lions and evolution. In P. P. G. Bateson and R. Hinde, eds., *Growing points in ethology*, pp. 281–301. Cambridge: Cambridge University Press.

———. 1985. Blood relatives: kin selection in a lion pride. In D. Macdonald, ed., *All the world's animals: Carnivores*, pp. 26–27. New York: Torstar.

Blaustein, A. R. and R. K. O'Hara. 1986. Kin recognition in tadpoles. *Scientific American* 254(1):108–16.

Bolwig, N. 1959. A study of nests built by mountain gorilla and chimpanzee. *S. Afr. J. Sci.* 55:286–91.

Buechner, H. K. 1971. Implications of social behavior in management of the Uganda kob. In *The behavior of ungulates and its relation to management*, vol. 2. Alberta, Canada: International Union for the Conservation of Nature and Natural Resources (IUCN) Conference, University of Calgary.

Busse, C. D. 1977. Chimpanzee predation as a possible factor in the evolution of red colobus monkey social organization. *Evolution* 31:907–11.

———. 1978. Do chimpanzees hunt cooperatively? *Am. Nat.* 112:767–70.

Butynski, T. M. 1982a. Vertebrate predation by primates: a review of hunting patterns and prey. *J. Hum. Evol.* 11:421–30.

———. 1982b. Harem-male replacement and infanticide in the blue monkey (*Cercopithecus mitis Stuhlmann*) in the Kibale Forest, Uganda. *Am. J. Primatology* 3:1–22.

———. 1984. Ecological survey of the Impenetrable (Bwindi) Forest, Uganda, and recommendations for its conservation and management. Unpublished report, New York: New York Zoological Society.

———1985. Primates and their conservation in the Impenetrable (Bwindi) Forest, Uganda. *Primate Conservation* (the newsletter and journal of the IUCN/SSC Primate Specialist Group) 6:68–72.

Bygott, J. D. 1972. Cannibalism among wild chimpanzees. *Nature* 238:410–11.

———. 1979. Agonistic behavior, dominance, and social structure of wild chimpanzees of the Gombe National Park. In Hamburg and McCown 1979, pp. 404–27.

References and Background Reading

Cahill, T. 1981. Gorilla tactics, life and love in gorilla country. *Geo* 3(12):100–16.

Cant, J. G. H. 1978. Ecology, locomotion, and social organization of spider monkeys. Ph.D. dissertation, University of California, Davis.

Caufield, C. 1985. *In the rain forest*. New York: Knopf.

Caughley, G. 1977. *Analysis of vertebrate populations*. New York: John Wiley.

Chivers, D. J., D. A. Wood, and A. Bilsborough, eds. 1984. *Food acquisition and processing in primates*. New York: Plenum.

Clark, C. B. 1977. A preliminary report on weaning among chimpanzees at the Gombe National Park, Tanzania. In S. Chevalier-Skolnikoff and F. E. Poirier, eds., *Primate bio-social development: biological, social, and ecological determinants*, pp. 235–60. New York: Garland.

Clutton-Brock, T. H. 1972. Feeding and ranging behavior of the red colobus monkey. Ph.D. dissertation, University of Cambridge.

Coley, P. D., J. P. Bryant, and F. S. Chapin III. 1985. Resource availability and plant antiherbivore defense. *Science* 230:895–99.

Cowles, J. T. 1937. Food tokens as incentives for learning by chimpanzees. *Psychol. Monogr.* 14:1–88.

Darwin, C. 1859. *On the origin of species*. London: John Murray.

———. 1871. *The descent of man, and selection in relation to sex*. London: John Murray.

Dawkins, R. 1976. *The selfish gene*. Oxford:Oxford University Press.

———. 1982. *The extended phenotype*. San Francisco: W. H. Freeman.

Diamond. J. M. 1975. The island delemma: lessons of modern biogeographical studies for the design of natural reserves. *Biological Conservation* 7(2): 129–46.

———. 1984. Making a chimp out of a man. *Discover* 5(12):54–60.

Donelson, J. E. and M. J. Turner. 1985. How the trypanosome changes it coat. *Scientific American* 252(2):44–51.

Douglas-Hamilton, I. and O. Douglas-Hamilton. 1975. *Among the elephants*. New York: Viking.

Eggeling, W. J. and I. R. Dale. 1951. *The indigenous trees of the Uganda Protectorate*. Uganda Protectorate: Government Printer.

Ehrlich, A. and P. Ehrlich. 1978. Hope for the tropical rain forests? *Mother Earth News* 53:110–11.

Eltringham, S. K. and R. C. Malpas. 1976. Elephant slaughter in Uganda. *Oryx* 13:334–35.

Emlen, J. M. 1966. The role of time and energy in food preference. *Am. Nat.* 100:611–17.

Endler, J. A. 1986. *Natural selection in the wild*. Princeton: Princeton University Press.

References and Background Reading

Ferster, C. B. 1964. Arithmetic behavior in chimpanzees. *Scientific American* 210(2):98–106.

Fleming, T. H. 1979. Do tropical frugivores compete for food? *Am. Zool.* 19: 1157–72.

Fossey, D. 1984. *Gorillas in the mist.* Boston: Houghton Mifflin.

Fouts, R. S. and R. L. Budd. 1979. Artificial and human language acquisition in the chimpanzee. In Hamburg and McCown 1979, pp. 374–92.

Frame, G. W. 1985. Cheetahs of the Serengeti: male-female differences in habitat exploitation. In D. Macdonald, ed., *All the world's animals: carnivores*, pp. 34–35. New York: Torstar.

Frame, L. H. and G. W. Frame. 1976. Female African wild dogs emigrate. *Nature* 263:227–29.

Freeland, W. J. and D. H. Janzen. 1974. Strategies in herbivory in mammals: the role of plant secondary compounds. *Am. Nat.* 108:269–89.

Frisch, J. E., S. J. 1968. Individual behavior and intertroop variability in Japanese macaques. in P. C. Jay, ed., *Primates studies in adaptation and variability*, pp. 243–52. San Francisco: Holt, Rinehart and Winston.

Gadgil, M. 1972. Male dimorphism as a consequence of sexual selection. *Am. Nat.* 106:574–80.

Galdikas, B. M. F. 1979. Orangutan adaptation at Tanjung Puting Reserve: mating and ecology. In Hamburg and McCown 1979, pp. 195–234.

_____. 1984. Adult female sociality among wild orangutans at Tanjung Puting Reserve. In M. F. Small, ed., *Female primates studies by women primatologists*, pp. 217–35. New York: Alan R. Liss.

Gardner, R. A. and B. T. Gardner. 1969. Teaching sign language to a chimpanzee. *Science* 165:664–72.

Gaulin, S. J. C. 1979. A Jarman/Bell model of primate feeding niches. *Human Ecology* 7(1):1–20.

Gause, G. F. 1934. *The struggle for existence.* Baltimore: Williams and Wilkins.

Ghiglieri, M. P. 1981. Wildlife log on the Omo River. *Oryx* 19(2):142–43.

_____. 1984a. *The chimpanzees of Kibale Forest.* New York: Columbia University Press.

_____. 1984b. The exploitation of patchy resources by chimpanzees in Kibale Forest, Uganda. In Rodman and Cant 1984, pp. 161–94.

_____. 1984c. The mountain gorilla: last of a vanishing tribe. *Mainstream* 15(3–4):36–40.

_____. 1985. The social ecology of chimpanzees. *Scientific American* 252(6): 102–13.

_____. 1986a. A river journey through Gunung Leuser National Park, Sumatra. *Oryx* 20(2): 104–10.

_____. 1986b. River of the red ape. *Mainstream* 17(4):29–33.

References and Background Reading

_____. 1987a. Toward a strategic model of hominid social evolution. In Heltne, ed., 1987, pp. 00–00.

_____. 1987b. War among the apes. *Discover* 8(10):00–00.

_____. 1988. Sociobiology of the great apes and the hominid ancestor. *J. Hum. Evol.* 17(3):000–000.

Ghiglieri, M. P., T. M. Butynski, T. T. Struhsaker, L. Leyland, S. J. Wallis, and P. Waser. 1982. Bush pig (*Potamochoerus porcus*) polychromatism and ecology in Kibale Forest, Uganda. *Afr. J. Ecol.* 20:231–36.

Gill, T. V. 1978. Conversing with Lana. In D. J. Chivers and J. Herbert, eds., *Recent advances in primatology*, pp. 861–66. San Francisco: Academic Press.

Goodall, A. G. 1977. Feeding and ranging behavior of a mountain gorilla group (*Gorilla gorilla berengei*) in the Tshibinda-Kahuzi region (Zaire). In T. H. Clutton-Brock, ed., *Primate ecology*, pp. 450–79. New York: Academic Press.

Goodall, J. [v.L.] 1963. My life among wild chimpanzees. *National Geographic* 124(2):272–308.

_____. 1964. Tool-using and aimed throwing in a community of free-living chimpanzees. *Nature* 201:1264–66.

_____. 1965a. New discoveries among Africa's chimpanzees. *National Geographic* 128(6):802–31.

_____. 1965b. Chimpanzees of the Gombe Stream Reserve. In I. DeVore, ed., *Primate behavior field studies of monkeys and apes*, pp. 425–73. San Francisco: Holt, Rinehart and Winston.

_____. 1968. The behaviour of free-living chimpanzees in the Gombe Stream Reserve. *Anim. Behav. Monogr.* 1:161–311.

_____. 1971a. Some aspects of aggressive behavior in a group of free-living chimpanzees. *Int. Soc. Sci. J.* 23:89–97.

_____. 1971b. *In the shadow of man*. Boston: Houghton Mifflin.

_____. 1973a. The behavior of chimpanzees in their natural habitat. *Am. J. Psychiatry* 130:1–12.

_____. 1973b. Cultural elements in a chimpanzee community. *Symp. IVth Int. Congr. Primat*, vol. 1, *Precultural primate behavior*, pp. 144–84.

_____. 1975. Chimpanzees of Gombe National Park: 13 years of research. In G. Kurth and I. Eibe-Eiblesfeldt, eds., *Hominisation und Verhalten*, pp. 74–106. Stuttgart: Fischer.

Goodall, J. 1977. Infant killing and cannibalism in free-living chimpanzees. *Folia Primatol.* 28:259–82.

_____. 1979. Life and death at Gombe. *National Geographic* 155(5):592–621.

_____. 1983. Population dynamics during a 15 year period in one community of free-living chimpanzees in the Gombe National Park, Tanzania. *Z. Tierpsychol.* 61:1–60.

_____. 1986. *The chimpanzees of Gombe.* Cambridge: Belknap Press of Harvard University Press.

Goodall, J., A. Bandoro, E. Bergman, C. Busse, H. Matama, E. Mpongo, A. Pierce, and D. Riss. 1979. Intercommunity interactions in the chimpanzee population of the Gombe National Park. In Hamburg and McCown 1979, pp. 13–54.

Gray, J. A. 1971. Sex differences in emotional behaviour in mammals including man. *Acta Psychologica* 35:29–46.

Hall, E. 1969. Hebb on hocus-pocus: a conversation. *Psychology Today* 31:20–28.

Halperin, S. D. 1979. Temporary association patterns in free-ranging chimpanzees. In Hamburg and McCown 1979, pp. 490–99.

Hamburg, D. A. 1986. New risks of prejudice, ethnocentrism, and violence. *Science* 231:533.

Hamburg, D. A. and E. R. McCown. 1979. *The great apes.* Menlo Park: Benjamin/Cummings.

Hamilton, W. D. 1964. The genetical theory of social behavior, I, II. *J. of Theoret. Biol.* 12:1–52.

_____. 1971. Geometry for the selfish herd. *J. of Theoret. Biol.* 31:295–311.

_____. 1972. Altruism and related phenomena mainly in social insects. *Ann. Rev. Ecol. System.* 3:193–232.

Hamilton, W. J. III and C. D. Busse. 1978. Primate carnivory and its significance to human diets. *BioScience* 28:761–66.

Handby, J. 1982. *Lion's share.* Boston: Houghton Mifflin.

Harako, R. 1981. The cultural ecology of hunting behavior among Mbuti Pygmies in the Ituri Forest, Zaïre. In R. S. O. Harding and G. Teleki, eds., *Omnivorous primates gathering and hunting in human evolution*, pp. 499–555. New York: Columbia University Press.

Harcourt, A. H. 1978. Strategies of immigration and transfer by primates, with particular reference to gorillas. *Z. Tierpsychol.* 48:401–20.

_____. 1979. The social relations and group structure of wild mountain gorillas. In Hamburg and McCown 1979, pp. 186–92.

_____. 1981. Intermale competition and the reproductive behavior of the great apes. In C. E. Graham, ed., *Reproductive biology of the great apes*, pp. 265–79. New York: Academic Press.

Harding, R. S. O. 1977. Patterns of movement in open-country baboons. *Am. J. Physical Anthropology* 47:349–54.

Hasegawa, T., M. Hiraiwa, T. Nishida, and H. Takasaki. 1983. New evidence on scavenging behavior in wild chimpanzees. *Current Anthropology* 24(2): 231–32.

Hausfater, G., J. Altmann, and S. Altmann. 1982. Long-term consistency of dominance relations among female baboons (*Papio cynocephalus*). *Science* 217:752–55.

References and Background Reading

Hausfater, G. and S. B. Hrdy, eds. 1984. *Infanticide: comparative and evolutionary perspectives*. New York: Aldine.

Heinlein, R. A. 1947. Jerry is a Man. In R. A. Heinlein, ed., *Assignment in eternity*, pp. 227–55. Reading, Pa.: Fantasy Press.

Heinrich, B. and G. A. Bartholomew. 1979. The ecology of African dung beetles. *Scientific American* 241(5):146–56.

Heltne, P. G., ed. 1987. *Understanding chimpanzees*. Chicago: Chicago Academy of Sciences.

Hladik, C. M. 1977. Chimpanzees of Gabon and chimpanzees of Gombe: some comparative data on the diet. In T. H. Clutton-Brock, ed., *Primate ecology*, pp. 481–501. London: Academic Press.

Hoogland, J. L. 1985. Infanticide in prairie dogs: females kill offspring of close kin. *Science* 230:1037–40.

Howe, H. F. 1980. Monkey dispersal and waste of neotropical fruit. *Ecology* 61(4):944–59.

Hrdy, S. B. 1977. Infanticide as a primary reproductive strategy. *American Scientist* 65:40–49.

———. 1979. Infanticide among animals: a review, classification, and examination of the implications for the reproductive strategies of females. *Ethology and Sociobiology* 1(1):13–40.

Hubbell, S. 1979. Tree dispersion, abundance, and diversity in a dry tropical forest. *Science* 203:1299–1309.

Hutchins, M. and D. P. Barash. 1976. Grooming in primates: implications for its utilitarian function. *Primates* 17(2):145–50.

Interagency Primate Steering Committee (Ad Hoc Task Force). 1980. *National Chimpanzee Breeding Program*. U.S. Department of Health and Human Services, National Institutes of Health, Bethesda, Md.

Isaac, G. 1978. The food sharing behavior of protohuman hominids. *Scientific American* 238(4):90–108.

Itani, J. 1980. Social structures of African great apes. *J. Reprod. Fert. Suppl.* 28:33–41.

Itani, J. and A. Suzuki. 1967. The social unit of chimpanzees. *Primates* 8:355–81.

Izawa, K. 1970. Unit groups of chimpanzees and their nomadism in the savanna woodland. *Primates* 11:1–46.

Izawa, K. and J. Itani. 1966. Chimpanzees in the Kasakati Basin, Tanganyika (I) Ecological study in the rainy season 1963–1964. *Kyoto Univ. African Studies* 1:73–156.

Janzen, D. H. 1979. How to be a fig. *Ann. Rev. Ecol. Syst.* 10:13–52.

Jolly, A. 1972. *The evolution of primate behavior*. New York: Macmillan.

Jones, C. and J. Sabater Pi. 1971. Comparative ecology of *Gorilla gorilla* (Savage and Wyman) and *Pan troglodytes* (Blumenbach) in Rio Muni, West Africa. *Bibliotheca Primatologica* 13:1–96.

Kamil, A. C. and T. D. Sargent, eds. 1981. *Foraging behavior: ecological, ethological, and physiological approaches.* New York: Garland.

Kano, T. 1971. The chimpanzees of Filabanga, Western Tanzania. *Primates* 12:229–46.

_____. 1972. Distribution and adaptation of the chimpanzee on the eastern shore of Lake Tanganyika. *Kyoto Univ. African Studies* 7:37–139.

_____. 1980. Social behavior of wild pygmy chimpanzees (*Pan paniscus*) of Wamba: a preliminary report. *J. Hum. Evol.* 9:243–60.

Kawabe, M. 1966. One observed case of hunting behavior among wild chimpanzees living near the savanna woodland of Western Tanzania. *Primates* 7:393–96.

Kawanaka, K. 1981. Infanticide and cannibalism in chimpanzees with special reference to the newly observed case in the Mahale Mountains. *African Studies Monographs* 1:69–99.

King, M. C. and A. C. Wilson. 1975. Evolution at two levels in humans and chimpanzees. *Science* 188:107–16.

Kingston, B. 1967. Working plan for Kibale and Itwara Central Forest Reserves. Entebbe, Uganda: Forestry Department.

Kohler, W. 1925. *The mentality of apes.* London: Routledge & Kegan Paul.

Kortlandt, A. 1962. Chimpanzees in the wild. *Scientific American* 206(5):128–38.

_____. 1983. Marginal habitats of chimpanzees. *J. Hum. Evol.* 12:231–78.

_____. 1984. Habitat richness, foraging range and diet in chimpanzees and some other primates. In D. J. Chivers, B. A. Wood, and A. Bilsborough, eds., *Food acquisition and processing in primates.* New York: Plenum.

Kortlandt, A. and J. C. J. Van Zon. 1969. The present state of research on the dehumanization hypothesis of African ape evolution. *Proc. 2nd Int. Congr. Primat.* 3:10–13.

Kruuk, H. 1972. *The spotted hyena.* Chicago: University of Chicago Press.

Kruuk, H. and M. Turner. 1967. Comparative notes on predation by lion, leopard, cheetah, and wild dog in the Serengeti area, East Africa. *Mammalia* 31:1–27.

Laitman, J. T. 1984. The anatomy of human speech. *Natural History* 93(8): 20–27.

Langdale-Brown, I., H. A. Osmaston, and J. G. Wilson. 1964. *The vegetation of Uganda and its bearing on land-use.* Entebbe, Uganda: Government Printer.

Lee, R. B. 1968. What hunters do for a living, or, how to make out on scarce resources. In R. B. Lee and I. DeVore, eds., *Man the hunter,* pp. 30–48. Chicago: Aldine.

Leopold, A. 1966. *A sand county almanac.* New York: Oxford University Press.

References and Background Reading

Lewin, R. 1982. How did humans evolve big brains? *Science* 216:840–41.

———. 1983a. Santa Rosalia was a goat. *Science* 221:636–39.

———. 1983b. Predators and hurricanes change ecology. *Science* 221:737–40.

———. 1983c. Is the orangutan a living fossil? *Science* 222:1222–23.

———. 1984. *Human evolution: an illustrated introduction*. New York: W. H. Freeman.

———. 1986a. Mass extinction without asteroids. *Science* 234:14–15.

———. 1986b. Damage to tropical forests, or why were there so many kinds of animals? *Science* 234:149–50.

MacArthur, R. H. and E. R. Pianka. 1966. On optimal use of a patchy environment. *Am. Nat.* 100:603–9.

Macdonald, D., ed. 1985. *All the world's animals: primates*. New York: Torstar.

McGinnis, P. R. 1979. Sexual behavior in free-living chimpanzees: consort relationships. In Hamburg and McCown 1979, pp. 428–39.

McGreal, S. 1987. World Health Organization (WHO) and Ecosystem Conservation Group (ECG) adopt Primate Specialist Group's "Policy statement on use of primates for biomedical purposes." *International Primate Protection League Newsletter* 4 May 1987, Attachment II.

McGrew, W. C. 1977. Socialization and object manipulation of wild chimpanzees. In S. Chevalier-Skolnikoff and F. E. Poirier, eds., *Primate bio-social development: biological, social, and ecological determinants*, pp. 261–88. New York: Garland.

———. 1979. Evolutionary implications of sex differences in chimpanzee predation and tool use. In Hamburg and McCown 1979, pp. 440–63.

———. 1981. The female chimpanzee as an evolutionary prototype. In F. Dahlberg, ed., *Woman the gatherer*, pp. 35–73. New Haven: Yale University Press.

McGrew, W. C., P. J. Baldwin, and C. E. G. Tutin. 1981. Chimpanzees in a hot, dry and open habitat: Mt. Assirik, Senegal, West Africa. *J. Hum. Evol.* 10:227–44.

McGrew, W. C., M. J. Sharman, P. J. Baldwin, and C. E. G. Tutin. 1982. On early hominid food niches. *Current Anthropology* 23:213–14.

McGrew, W. C. and C. E. G. Tutin. 1978. Evidence for a social custom in wild chimpanzees? *Man* 13:234–51.

Mack, D. and R. A. Mittermeier. 1985. *The international primate trade vol. 1. Legislation, trade, and captive breeding*. Washington, D.C.: WWF-US, TRAFFIC (U.S.A.) and IUCN/SSC Primate Specialist Group.

MacKinnon, J. 1974a. The behaviour and ecology of wild orangutans *Pongo pygmaeus*. *Anim. Behav.* 22:3–74.

———. 1974b. *In search of the red ape*. New York: Holt, Rinehart and Winston.

———. 1979. Reproductive behavior in wild orangutan populations. In Hamburg and McCown 1979, pp. 256–73.

McNab, B. 1963. Bioenergetics and the determination of home range size. *Am. Nat.* 97:133–40.

Marler, P. 1969. Vocalizations of wild chimpanzees. *Proc. 2nd Int. Congr. Primat.* 1:94–100.

Marler, P. and L. Hobbet. 1975. Individuality in a long-range vocalization of wild chimpanzees. Z. *Tierpsychol* 38:97–109.

Martin, P. S. and R. G. Klein, eds. 1984. *Quaternary extinctions, a prehistoric revolution.* Tucson: University of Arizona Press.

Maynard Smith, J. 1978. The evolution of behavior. *Scientific American* 239(3):176–92.

———. 1982. *Evolution and the theory of games.* Cambridge: Cambridge University Press.

Mech, L. D. 1970. *The wolf: the ecology and behavior of an endangered species.* New York: Natural History Press.

———. 1977. Productivity, mortality, and population trends of wolves in northeastern Minnesota. *J. Mammal.* 58:559–74.

Menzel, E. W. 1973. Chimpanzee spatial memory organization. *Science* 182:943–45.

———. 1979. Communication of object locations in a group of young chimpanzees. In Hamburg and McCown 1979, pp. 358–71.

Merfield, F. G. and H. Miller. 1956. *Gorillas were my neighbors.* London: Longman.

Miller, C. 1975. *The lunatic express.* New York: Macmillan.

Miller, D. A. 1977. Evolution of primate chromosomes. *Science* 198:1116–24.

Miller. R. S. 1967. Pattern and process in competition. *Adv. Ecol. Res.* 4:1–74.

Milton, K. and M. L. May. 1976. Body weight, diet and home range area in primates. *Nature* 259:459–62.

Mitani, J. C. 1985. Mating behavior of adult male orangutans in the Kutai Game Reserve, Indonesia. *Anim. Behav.* 33:392–402.

Moore, J. 1985. Chimpanzee survey in Mali, West Africa. *Primate Conservation* (the newsletter and journal of the IUCN/SSC Primate Specialist Group) 6:59–63.

Moorehead, A. 1962. *The Blue Nile.* New York: Harper and Row.

Morgan, C. J. 1979. Eskimo hunting groups, social kinship and the possibility of kin selection among humans. *Ethology and Sociobiology* 1(1):83–86.

Morris, K. and J. Goodall. 1977. Competition for meat between chimpanzees and baboons of the Gombe National Park. *Folia Primatol.* 28:109–21.

Mounin, G. 1976. Language, communication, chimpanzees. *Current Anthropology* 17(1):1–21.

References and Background Reading

Murdock, G. P. 1967. Ethnographic Atlas: a summary. *Ethnography* 6:109–36.

Myers, N. 1979. *The sinking ark*. New York: Pergamon.

_____. 1984. *The primary source*. New York: W. W. Norton.

Nicholson, N. A. 1977. A comparison of early behavioral development in wild and captive chimpanzees. In S. Chevalier-Skolnikoff and F. E. Poirier, eds., *Primate bio-social development: biological, social, and ecological determinants*, pp. 539–60. New York: Garland.

Nishida, T. 1968. The social group of wild chimpanzees in the Mahali Mountains. *Primates* 9:167–224.

_____. 1970. Social behavior and relationships among wild chimpanzees of the Mahali Mountains. *Primates* 11:47–87.

_____. 1973. The ant-gathering behavior by the use of tools among wild chimpanzees of the Mahali Mountains. *J. Hum. Evol.* 2:357–70.

_____. 1976. The bark-eating habits in primates, with special reference to their status in the diet of wild chimpanzees. *Folia Primatol.* 25:277–87.

_____. 1979. The social structure of chimpanzees of the Mahale Mountains. In Hamburg and McCown 1979, pp. 73–122.

_____. 1980. Local differences in response to water among wild chimpanzees. *Folia Primatol.* 33:189–209.

_____. 1983. Alpha status and agonistic alliance in wild chimpanzees (*Pan troglodytes schweinfurthii*). *Primates* 24(3):318–36.

Nishida, T. and M. Hiraiwa-Hasegawa. 1985. Responses to a mother-son pair in the wild chimpanzee: a case report. *Primates* 26(1):1–13.

Nishida, T., M. Hiraiwa-Hasegawa, T. Hasegawa, and Y. Takahata. 1985. Group extinction and female transfer in wild chimpanzees in the Mahale National Park, Tanzania. *Z. Tierpsychol.* 67:284–301.

Nishida, T. and K. Kawanaka. 1972. Inter-unit-group relationships among wild chimpanzees of the Mahali Mountains. *Kyoto Univ. African Studies* 7: 131–69.

_____. 1985. Within group cannibalism by adult male chimpanzees. *Primates* 26(3):274–84.

Nishida, T. and S. Uehara. 1981. Kitongwe name of plants: a preliminary listing. *African Studies Monographs* 1:109–31.

_____. 1983. Natural diet of chimpanzees (*Pan troglodytes schweinfurthii*): long-term record from the Mahale Mountains, Tanzania. *African Studies Monographs* 3:109–30.

Nishida, T., R. W. Wrangham, J. Goodall, and S. Uehara. 1983. Local differences in plant-feeding habits of chimpanzees between the Mahale Mountains and Gombe National Park, Tanzania. *J. Hum. Evol.* 12:467–80.

Nissen, H. W. 1931. A field study of the chimpanzee. *Comparative Psychology Monographs* 8(1):1–122.

References and Background Reading

Oates, J. F. 1974. The ecology and behavior of black and white colobus (*Colobus guereza Ruppel*) in East Africa. Ph.D. dissertation, University of London.

Owens, M. and D. Owens. 1984. *Cry of the Kalahari*. Boston: Houghton Mifflin.

Packer, C. 1975. Male transfer in olive baboons. *Nature* 255:219–20.

_____. 1977. Reciprocal altruism in *Papio anubis*. *Nature* 265:441–43.

Patterson, F. and E. Linden. 1981. *The education of Koko*. New York: Holt, Rinehart and Winston.

Perry, D. R. 1984. The canopy of the tropical rain forest. *Scientific American* 251(5):138–47.

Pierce, A. H. 1978. Ranging patterns and associations of a small community of chimpanzees in Gombe National Park, Tanzania. In D. J. Chivers and J. Herbert, eds., *Recent advances in primatology vol. 1, Behavior*, pp. 59–61. San Francisco: Academic Press.

Platt, J. R. 1964. Strong inference. *Science* 146:347–53.

Pooley, A. C. and C. Gans. 1976. The Nile crocodile. *Scientific American* 234(4):114–24.

Premack, A. J. and D. Premack. 1972. Teaching sign language to an ape. *Scientific American* 227(4):92–99.

Premack, D. and A. J. Premack. 1983. *The mind of an ape*. New York: W. W. Norton.

Preuschoft, H., D. Chivers, W. Brockelman, and N. Creel, eds. 1985. *The lesser apes evolutionary and behavioural ecology*. Edinburgh: Edinburgh University Press.

Pusey, A. E. 1977. Physical and social development of wild chimpanzees. Ph.D dissertation, Stanford University.

_____. 1979. Intercommunity transfer of chimpanzees in Gombe National Park. In Hamburg and McCown 1979, pp. 465–80.

_____. 1980. Inbreeding avoidance in chimpanzees. *Anim. Behav.* 28:543–52.

Pusey, A. E. and C. Packer. 1983. Once and future kings. *Natural History* 92(8):54–63.

Pyke, G. H., H. R. Pulliam, and E. L. Charnov. 1977. Optimal foraging: a selective review of theory and tests. *Ann. Rev. Biol.* 52:137–54.

Raeburn, P. 1983. An uncommon chimp. *Science 83* 4(5):40–48.

Rahm, U. 1967. Observations during chimpanzee captures in the Congo. In D. Starck, R. Schneider, and H. J. Kuhn, eds., *Progress in primatology: first congress of the Int. Primat. Soc.*, pp. 195–206. Stuttgart: Gustav Fischer.

Redmond, I. 1986. Law of the Jungle. Special Report of the International Primate Protection League. P.O. Box 766, Summerville, NC 29484.

Reynolds, V. 1963. An outline of the behavior and social organization of forest-living chimpanzees. *Folia Primatol.* 1:95–102.

References and Background Reading

_____. 1965. *Budongo*. Garden City, N.Y.: Natural History Press.

_____. 1975. How wild are the Gombe chimpanzees? *Man* 10:123–25.

Reynolds, V. and F. Reynolds. 1965. Chimpanzees in the Budongo Forest. In I. DeVore, ed., *Primate behavior field studies of monkeys and apes*, pp. 368–424. San Francisco: Holt, Rinehart and Winston.

Richard, A. 1985. *Primates in nature*. San Francisco: W. H. Freeman.

Richards, P. W. 1952. *The tropical rain forest*. London: Cambridge University Press.

Rijksen, H. D. 1978. A *field study on Sumatran orangutans (Pongo pygmaeus abelii Lessen 1927): ecology, behavior and conservation*. Wageningen, the Netherlands: H. Veenmen and Zonen, B. V.

_____. 1982. How to save the mysterious 'man of the forest'? In L. E. M. de Boer, ed., *The orangutan: its biology and conservation*, pp. 468–524. The Hague: Dr. W. Junk.

Riss, D. C. and C. R. Busse. 1977. Fifty-day observation of a free-ranging adult male chimpanzee. *Folia Primatol.* 28:283–97.

Riss, D. C. and J. Goodall. 1977. The recent rise to alpha-rank in a population of free-living chimpanzees. *Folia Primatol.* 27:134–51.

Robbins, D. and C. R. Bush. 1973. Memory in great apes. *J. Exp. Psychol.* 97:344–48.

Rodman, P. S. 1973. Population composition and adaptive organization among orangutans of the Kutai Reserve. In R. P. Michael and J. H. Crook, eds., *Ecology and behavior of primates*, pp. 171–209. London: Academic Press.

_____. 1979. Individual activity patterns and the solitary nature of orangutans. In Hamburg and McCown 1979, pp. 235–56.

_____. 1984. Foraging and social systems of orangutans and chimpanzees. In Rodman and Cant 1984, pp. 134–60.

Rodman, P. S. and J. G. H. Cant, eds. 1984. *Adaptations for foraging in nonhuman primates*. New York: Columbia University Press.

Rosenthal, G. A. 1986. The chemical defenses of higher plants. *Scientific American* 254(1):94–99.

Roush, G. J. 1982. On saving diversity. *The Nature Conservancy News* 32(1):4–10.

Rowell, T. E. 1966. Forest living baboons in Uganda. *J. of Zool. London* 149:344–64.

Rudnai, J. 1973. Reproductive biology of lions (*Panthera leo massaica* Neuman) in Nairobi National Park. *E. Afr. Wildl. J.* 11:241–53.

Rudran, R. 1976. The socio-ecology of the blue monkey (*Cercopithecus mitis Stuhlmanni*) of the Kibale Forest, Uganda. Ph.D. dissertation, University of Maryland.

Rumbaugh, D. M. 1970. Learning skills of anthropoids. In L. A. Rosenblem, ed., *Primate Behavior 1*, pp. 1–70. London: Academic Press.

_____. 1974. Lana's world. *Yerkes Newsletter* 11:2–7.

References and Background Reading

Sagan, C. and A. Druyan. 1985. *Comet*. New York: Random House.

Sarich, V. M. and J. E. Cronin. 1976. Molecular systematics of the primates. In M. Goodman and R. Tashian, eds., *Molecular anthropology*, pp. 141–70. New York: Plenum.

Sayfarth, R. M. 1976. Social relationships among adult female baboons. *Anim. Behav.* 24:917–28.

Scientific American staff. 1986. Seeing the forest. *Scientific American* 255 (3):66–67.

Schaller, G. B. 1961. The orang-utan in Sarawak. *Zoologica* 46:73–82.

———. 1963. *The mountain gorilla*. Chicago: University of Chicago Press.

———. 1964. *The year of the gorilla*. Chicago: University of Chicago Press.

———. 1972. *The Serengeti lion*. Chicago: University of Chicago Press.

Schneirla, T. C. 1971. *Army ants: a study in social organization*, edited by H. R. Topoff. San Francisco: W. H. Freeman.

Schoener, T. 1971. Theory of feeding strategies. *Ann. Rev. Ecol. System.* 2: 369–40.

Shipman, P. 1986. Scavenging or hunting in early hominids: theoretical framework and tests. *American Anthropologist* 88(1): 27–43.

Short, R. V. 1979. Sexual selection and its component parts, somatic and genital selection as illustrated by man and the great apes. *Advances in the Study of Behavior* 9:131–58.

Sibley, C. G. and J. E. Ahlquist. 1984. The phylogeny of the hominoid primates, as indicated by DNA-DNA hybridization. *Journal of Molecular Evolution* 20:2–15.

———. 1986. Reconstructing bird phylogeny by comparing DNA's. *Scientific American* 254(3):82–92.

Silberbauer, G. 1981. Hunter/gatherers of the central Kalahari. In R. S. O. Harding and G. Teleki, eds. *Omnivorous primates gathering and hunting in human evolution*, pp. 455–98. New York: Colombia University Press.

Silk, J. B. 1978. Patterns of food sharing among mother and infant chimpanzees at Gombe National Park, Tanzania. *Folia Primatol.* 29:129–41.

Simpson, M. J. A. 1973. The social grooming of male chimpanzees. In R. P. Michael and J. H. Crook, eds., *Comparative ecology and behaviour of primates*, pp. 411–505. London: Academic Press.

Skorupa, J. P. 1986. Responses of rainforest primates to selective logging in Kibale Forest, Uganda: a summary report. In K. Benirschke, ed., *Primates, the road to self-sustaining populations*, pp. 57–70. New York: Springer-Verlag.

Sparks, J. 1967. Allogrooming in primates: a review. In D. Morris, ed., *Primate ethology*, pp. 190–225. Garden City, N.Y.: Doubleday.

Stebbins, G. L. and F. J. Ayala. 1985. The evolution of Darwinism. *Scientific American* 253(1):72–82.

Stephenson, C. 1938. Leiningen versus the ants. In *The most dangerous game*, pp. 30–53. New York: Berkley (1957).

Struhsaker, T. T. 1967. Social structure among vervet monkeys (*Cercopithecus aethiops*). *Behaviour* 29:83–121.

_____. 1975. *The red colobus monkey*. Chicago: University of Chicago Press.

_____. 1977. Infanticide and social organization in the redtail monkey (*Cercopithecus ascanius schmidti*) in the Kibale Forest, Uganda. *Z. Tierpsychol.* 45:75–84.

_____. 1978. Food habits of five monkey species in the Kibale Forest, Uganda. In D. J. Chivers and J. Herbert, eds., *Recent advances in primatology, vol. 1, behavior*, pp. 225–48. London: Academic Press.

_____. 1982. Polyspecific associations among tropical rain forest primates. *Z. Tierpsychol.* 57:268–304.

_____. 1985. Hybrid monkeys of the Kibale Forest. In D. Macdonald 1985, pp. 100–101.

Struhsaker, T. T. and P. Hunkeler. 1971. Evidence of tool-using by chimpanzees in the Ivory Coast. *Folia Primatol.* 15:212–19.

Strum, S. C. 1975. Life with the Pumphouse Gang. *National Geographic* 147(5):672–91.

_____. 1981. Processes and products of change: baboon predatory behavior at Gilgil, Kenya. In R. S. O. Harding and G. Teleki, eds., *Omnivorous primates gathering and hunting in human evolution*, pp. 255–302. New York: Columbia University Press.

Sugiyama, Y. 1968. Social organization of chimpanzees in the Budongo Forest, Uganda. *Primates* 9:225–58.

_____. 1969. Social behavior of chimpanzees in the Budongo Forest, Uganda. *Primates* 10:197–225.

_____. 1972. Social characteristics and socialization of wild chimpanzees. In F. E. Poirier, ed., *Primate socialization*, pp. 145–63. New York: Random House.

Sugiyama, Y. and J. Koman. 1979. Social structure and dynamics of wild chimpanzees at Bossou, Guinea. *Primates* 20(3):323–39.

Sugiyama, Y. and H. Ohsawa. 1982. Population dynamics of Japanese monkeys with special references to the effect of artificial feeding. *Folia Primatol.* 39:238–63.

Susman, R. L. ed. 1984. *The pygmy chimpanzee evolutionary biology and behavior*. New York: Plenum.

Suzuki, A. 1966. On the insect-eating habits among wild chimpanzees living in the savanna woodland of western Tanzania. *Primates* 7:481–87.

_____. 1969. An ecological study of chimpanzees in a savanna woodland. *Primates* 10:103–48.

_____. 1971. Carnivory and cannibalism observed in forest-living chimpanzees. *J. Anthrop. Soc. Nippon* 74:30–48.

References and Background Reading

Tahoe Daily Tribune. 1984. Chetta lives out the golden years. Dec. 7, 1984, p. 2c.

Takahata, Y., T. Hasegawa, and T. Nishida. 1984. Chimpanzee predation in the Mahale Mountains from August 1979 to May 1982. *Int. J. Primatology* 5(3):213–33.

Tanner, N. M. 1986. Gathering by females: the chimpanzee model revisited and the gathering hypothesis. In W. H. Kinzey, ed., *The evolution of primate behavior: primate models*, pp. 3–27. Albany: State University of New York Press.

Teleki, G. 1972. The omnivorous chimpanzee. *Scientific American* 228(1):33–42.

———. 1973a. Group response to the accidental death of a chimpanzee in Gombe National Park, Tanzania. *Folia Primatol.* 20:81–94.

———. 1973b. Notes on chimpanzee interactions with small carnivores in Gombe National Park, Tanzania. *Primates* 14:407–11.

———. 1981. The omnivorous diet and eclectic feeding habits of chimpanzees in Gombe National Park, Tanzania. In R. S. O. Harding and G. Teleki, eds., *Omnivorous primates gathering and hunting in human evolution*, pp. 303–43. New York: Columbia University Press.

———. 1987. The once "common" chimpanzee (*Pan troglodytes*) of Equatorial Africa: continuing exploitation and decimation of a sibling species. In Heltne 1987.

Terrace, H. S. 1979. How Nim Chimpsky changed my mind. *Psychology Today* 13(6):65–76.

Thomas, E. M. 1959. *The harmless people*. New York: Knopf.

Topoff, H. R. 1972. The social behavior of army ants. *Scientific American* 227(5):70–79.

Trivers, R. L. 1971. The evolution of reciprocal altruism. *Quart. Rev. Biol.* 46:35–57.

———. 1972. Parental investment and sexual selection. In B. Campbell, ed., *Sexual selection and the descent of man, 1871–1971*, pp. 136–79. Chicago: Aldine Atherton.

———. 1974. Parent-offspring conflict. *Am. Zool.* 12:249–64.

Turnbull, C. 1961. *The forest people: a study of the pygmies of the Congo*. New York: Simon and Schuster.

Turton, D. 1978. Territorial organization and age among the Mursi. In P. W. Baxter and U. Almagor, eds., *Age, generation and time*, pp. 95–131. London: Hurst.

Tutin, C. E. G. 1975. Exceptions to promiscuity in a feral chimpanzee community. In S. Kondo, M. Kawai, and A. Ehara, eds., *Contemporary primatology*, pp. 445–49. Basel: S. Karger.

———. 1979. Mating patterns and reproductive strategies in a community of wild chimpanzees (*Pan troglodytes schweinfurthii*). *Behav. Ecol. Sociobiol.* 6:29–38.

_____. 1980. Reproductive behaviour of wild chimpanzees in the Gombe National Park, Tanzania. *J. Reprod. Fert. Suppl.* 28:43–57.

Tutin, C. E. G. and M. Fernandez. 1984. Nationwide census of Gorilla (*Gorilla g. gorilla*) and chimpanzee (*Pan t. troglodytes*) populations in Gabon. *Am. J. Primatology* 6:313–36.

Tutin, C. E. G. and P. R. McGinnis. 1981. Chimpanzee reproduction in the wild. In C. E. Graham, ed., *Reproductive biology of the great apes*, pp. 239–64. London: Academic Press.

Vail, L. 1977. Ecology and history: the example in eastern Zambia. *J. of Southern African Studies* 3:130–55.

Van Orsdol, K. G. 1979. Slaughter of the innocents. *Animal Kingdom* 82(6): 19–26.

_____. 1985. Lion. In Macdonald, ed., 1985, pp. 20–25.

_____. 1986. Agricultural encroachment in Uganda's Kibale Forest. *Oryx* 20(2): 115–17.

Vayda, A. P. 1976. *War in ecological perspective: persistence, change and adaptive process in three oceanic societies.* New York: Plenum.

Waal, F. B. M. de. 1984. *Chimpanzee politics power & sex among the apes.* New York: Harper Colophon.

Washburn, S. L. and I. DeVore. 1961. Social behavior of baboons and early man. In S. L. Washburn, ed., *Social life of early man*, pp. 91–105. Viking Fund Publications in Anthropology, no. 31. Wenner Gren Foundation.

_____. 1961. The social life of baboons. *Scientific American* 204(6):62–71.

Watts, D. P. 1985. Relations between group size and composition and feeding competition in mountain gorilla groups. *Anim. Behav.* 33:72–85

Westoby, M. 1974. An analysis of diet selection by large generalist herbivores. *Am. Nat.* 108:290–304.

White, F. J. 1986. Behavioral ecology of the pygmy chimpanzee. Ph.D. dissertation, State University of New York, Stony Brook.

Wiebes, J. T. 1979. Co-evolution of figs and their insect pollinators. *Ann. Rev. Ecol. Syst.* 10:1–12.

Wiley, R. H. and D. G. Richards. 1978. Physical constraints on acoustic communication in the atmosphere: implications for the evolution of animal vocalizations. *Behavioral Ecology and Sociobiology* 3(1):69–94.

Williams, G. C. 1966. *Adaptation and natural selection: a critique of some current evolutionary thought.* Princeton, N.J.: Princeton University Press.

Wilson, A. C. 1985. The molecular basis of evolution. *Scientific American* 253(4):164–73.

Wilson, E. O. 1971. *The insect societies.* Cambridge, Mass.: Belknap Press of Harvard University Press.

_____. 1975. *Sociobiology, the new synthesis.* Cambridge, Mass.: Belknap Press of Harvard University Press.

References and Background Reading

————. 1978. *On human nature.* Cambridge, Mass.: Harvard University Press.

————. 1985. Time to revive systematics. *Science* 230:1227.

Wing, L. D. and I. O. Buss. 1970. Elephants and forests. *Wildlf. Monogr.*, no. 19.

Wolf, K. and S. T. Schulman. 1984. Male response to "stranger" females as a function of female reproductive value among chimpanzees. *Am. Nat.* 123:163–74.

Wolfe, J. B. 1936. Effectiveness of token rewards for chimpanzees. *Comp. Psychol. Monogr.* 12:1–72.

Wolfheim, J. H. 1983. *Primates of the world: distribution, abundance, and conservation.* Seattle: University of Washington Press.

Woodburn, J. 1968. An introduction to Hadza ecology. In R. B. Lee and I. DeVore, eds., *Man the hunter*, pp. 49–55. Chicago: Aldine.

Wrangham, R. W. 1974. Artifical feeding of chimpanzees and baboons in their natural habitats. *Anim. Behav.* 22:83–93.

————. 1975. The behavioural ecology of chimpanzees in Gombe National Park, Tanzania. Ph.D. dissertation, University of Cambridge.

————. 1977. Feeding behaviour of chimpanzees in Gombe National Park, Tanzania. In T. H. Clutton-Brock, ed., *Primate ecology*, pp. 504–38. London: Academic Press.

————. 1979a. On the evolution of ape social systems. *Social Science Information* 18:335–68.

————. 1979b. Sex differences in chimpanzee dispersion. In Hamburg and McCown 1979, pp. 481–90.

————. 1980. An ecological model of female-bonded primate groups. *Behaviour* 75:262–300.

————. 1986. The significance of African apes for reconstructing human social evolution. In W. G. Kinzey, ed., *The evolution of human behavior: primate models*, pp. 51–71. Albany: State University of New York.

Wrangham, R. W. and T. Nishida. 1983. *Aspilia* spp. leaves: a puzzle in the feeding behavior of wild chimpanzees. *Primates* 24(2):276–82.

Wrangham, R. W. and B. B. Smuts. 1980. Sex differences in the behavioural ecology of chimpanzees in the Gombe National Park, Tanzania. *J. Reprod. Fert., Suppl.* 28:13–31.

Wu, W. M. H., W. G. Holmes, S. R. Medina, and G. P. Sackett. 1980. Kin preference in infant *Macaca nemistrina. Nature* 285:225–27.

Yengoyan, A. A. 1968. Demographic and ecological influences on aboriginal Australian marriage sections. In R. B. Lee and I. DeVore, eds., *Man the hunter*, pp. 185–99. Chicago: Aldine.

Index

Index

Index

Index

Index

Index

309

Index

310

Index

Index

Index

Index